T0305733

Linear Models with
Python

CHAPMAN & HALL/CRC
Texts in Statistical Science Series

Joseph K. Blitzstein, *Harvard University, USA*
Julian J. Faraway, *University of Bath, UK*
Martin Tanner, *Northwestern University, USA*
Jim Zidek, *University of British Columbia, Canada*

Recently Published Titles

Linear Models with Python
Julian J. Faraway

Introduction to Probability, Second Edition
Joseph K. Blitzstein and Jessica Hwang

Theory of Spatial Statistics
A Concise Introduction
M.N.M van Lieshout

Bayesian Statistical Methods
Brian J. Reich and Sujit K. Ghosh

Sampling
Design and Analysis, Second Edition
Sharon L. Lohr

The Analysis of Time Series
An Introduction with R, Seventh Edition
Chris Chatfield and Haipeng Xing

Time Series
A Data Analysis Approach Using R
Robert H. Shumway and David S. Stoffer

Practical Multivariate Analysis, Sixth Edition
Abdelmonem Afifi, Susanne May, Robin A. Donatello, and Virginia A. Clark

Time Series: A First Course with Bootstrap Starter
Tucker S. McElroy and Dimitris N. Politis

Probability and Bayesian Modeling
Jim Albert and Jingchen Hu

Surrogates
Gaussian Process Modeling, Design, and Optimization for the Applied Sciences
Robert B. Gramacy

For more information about this series, please visit:
https://www.crcpress.com/Chapman--HallCRC-Texts-in-Statistical-Science/book-series/CHTEXSTASCI

Linear Models with Python

Julian J. Faraway

CRC Press
Taylor & Francis Group
Boca Raton London New York

CRC Press is an imprint of the
Taylor & Francis Group, an **informa** business

A CHAPMAN & HALL BOOK

First edition published 2021
by CRC Press
6000 Broken Sound Parkway NW, Suite 300, Boca Raton, FL 33487-2742

and by CRC Press
2 Park Square, Milton Park, Abingdon, Oxon, OX14 4RN

Library of Congress Cataloging-in-Publication Data

Names: Faraway, Julian James, author.
Title: Linear models with Python / Julian J. Faraway.
Description: First edition. | Boca Raton : CRC Press, 2021. | Series:
Chapman & Hall/CRC texts in statistical science | Includes
bibliographical references and index.
Identifiers: LCCN 2020038706 | ISBN 9781138483958 (hardback) | ISBN
9781351053419 (ebook)
Subjects: LCSH: Linear models (Statistics) | Python (Computer program
language)
Classification: LCC QA279.F3695 2021 | DDC 519.50285/5133--dc23
LC record available at https://lccn.loc.gov/2020038706

ISBN: 978-1-138-48395-8 (hbk)
ISBN: 978-1-351-05341-9 (ebk)

Contents

Preface   ix

1   **Introduction**   **1**
   1.1   Before You Start   1
   1.2   Initial Data Analysis   2
   1.3   When to Use Linear Modeling   6
   1.4   History   7

2   **Estimation**   **15**
   2.1   Linear Model   15
   2.2   Matrix Representation   16
   2.3   Estimating β 17
   2.4   Least Squares Estimation   18
   2.5   Examples of Calculating $\hat{\beta}$ 19
   2.6   Example   19
   2.7   Computing Least Squares Estimates   22
   2.8   Gauss–Markov Theorem   24
   2.9   Goodness of Fit   26
   2.10   Identifiability   28
   2.11   Orthogonality   31

3   **Inference**   **37**
   3.1   Hypothesis Tests to Compare Models   37
   3.2   Testing Examples   39
   3.3   Permutation Tests   44
   3.4   Sampling   45
   3.5   Confidence Intervals for β 47
   3.6   Bootstrap Confidence Intervals   48

4   **Prediction**   **53**
   4.1   Confidence Intervals for Predictions   53
   4.2   Predicting Body Fat   54
   4.3   Autoregression   56
   4.4   What Can Go Wrong with Predictions?   58

5   Explanation   **61**
   5.1   Simple Meaning   61
   5.2   Causality   63
   5.3   Designed Experiments   64
   5.4   Observational Data   65
   5.5   Matching   67
   5.6   Covariate Adjustment   70
   5.7   Qualitative Support for Causation   71

6   Diagnostics   **75**
   6.1   Checking Error Assumptions   75
      6.1.1   Constant Variance   75
      6.1.2   Normality   80
      6.1.3   Correlated Errors   83
   6.2   Finding Unusual Observations   85
      6.2.1   Leverage   85
      6.2.2   Outliers   87
      6.2.3   Influential Observations   91
   6.3   Checking the Structure of the Model   93
   6.4   Discussion   96

7   Problems with the Predictors   **101**
   7.1   Errors in the Predictors   101
   7.2   Changes of Scale   105
   7.3   Collinearity   108

8   Problems with the Error   **115**
   8.1   Generalized Least Squares   115
   8.2   Weighted Least Squares   117
   8.3   Testing for Lack of Fit   121
   8.4   Robust Regression   125
      8.4.1   M-Estimation   125
      8.4.2   High Breakdown Estimators   128

9   Transformation   **135**
   9.1   Transforming the Response   135
   9.2   Transforming the Predictors   140
   9.3   Broken Stick Regression   140
   9.4   Polynomials   142
   9.5   Splines   148
   9.6   Additive Models   150
   9.7   More Complex Models   152

10 Model Selection   **155**
  10.1 Hierarchical Models   156
  10.2 Hypothesis Testing-Based Procedures   156
  10.3 Criterion-Based Procedures   160
  10.4 Sample Splitting   163
  10.5 Crossvalidation   167
  10.6 Summary   169

11 Shrinkage Methods   **173**
  11.1 Principal Components   173
  11.2 Partial Least Squares   184
  11.3 Ridge Regression   187
  11.4 Lasso   191
  11.5 Other Methods   194

12 Insurance Redlining — A Complete Example   **197**
  12.1 Ecological Correlation   197
  12.2 Initial Data Analysis   199
  12.3 Full Model and Diagnostics   202
  12.4 Sensitivity Analysis   204
  12.5 Discussion   207

13 Missing Data   **211**
  13.1 Types of Missing Data   211
  13.2 Representation and Detection of Missing Values   212
  13.3 Deletion   213
  13.4 Single Imputation   215
  13.5 Multiple Imputation   217
  13.6 Discussion   219

14 Categorical Predictors   **221**
  14.1 A Two-Level Factor   221
  14.2 Factors and Quantitative Predictors   225
  14.3 Interpretation with Interaction Terms   228
  14.4 Factors with More Than Two Levels   230
  14.5 Alternative Codings of Qualitative Predictors   235

15 One-Factor Models   **241**
  15.1 The Model   241
  15.2 An Example   242
  15.3 Diagnostics   245
  15.4 Pairwise Comparisons   246
  15.5 False Discovery Rate   248

16 Models with Several Factors   **253**
    16.1 Two Factors with No Replication   253
    16.2 Two Factors with Replication   257
    16.3 Two Factors with an Interaction   262
    16.4 Larger Factorial Experiments   266

17 Experiments with Blocks   **273**
    17.1 Randomized Block Design   274
    17.2 Latin Squares   278
    17.3 Balanced Incomplete Block Design   282

A About Python   **289**

Bibliography   **291**

Index   **295**

Preface

This is a book about linear models in statistics. A linear model describes a quantitative response in terms of a linear combination of predictors. You can use a linear model to make predictions or explain the relationship between the response and the predictors. Linear models are very flexible and widely used in applications in physical science, engineering, social science and business. Linear models are part of the core of statistics and understanding them well is crucial to a broader competence in the practice of statistics.

This is not an introductory textbook. You will need some basic prior knowledge of statistics as might be obtained in one or two courses at the university level. You will need to be familiar with essential ideas such as hypothesis testing, confidence intervals, likelihood and parameter estimation. You will also need to be competent in the mathematical methods of calculus and linear algebra. This is not a particularly theoretical book, as I have preferred intuition over rigorous proof. Nevertheless, successful statistics requires an appreciation of the principles. It is my hope that the reader will absorb these through the many examples I present.

This book is written in three languages: English, Mathematics and Python. I aim to combine these three seamlessly to allow coherent exposition of the practice of linear modeling. This requires the reader to become somewhat fluent in Python. This is not a book about learning Python but like any foreign language, one becomes proficient by practicing it rather than by memorizing the dictionary. The reader is advised to look elsewhere for a basic introduction to Python, but should not hesitate to dive into this book and pick it up as you go. I shall try to help. See the Appendix to get started.

The book's website can be found at:

https://julianfaraway.github.io/LMP/

This book has an ancestor: Faraway (2014) entitled *Linear Models with R*. Clearly, the book you hold now is about Python and not R but it is not an exact translation. Although I was able to accomplish almost all of the R book in this Python book, I found reason for variation:

1. Python and R are similar (at least in the way they are used for statistics) but they make different things easy and difficult. Hence, it is natural to flow along the Python path for easier ways to accomplish the same tasks.

2. Python is multi-talented, but R was designed to do statistics. R has a very large library of packages for statistical methods while Python has few. This has restricted

the choice of methods I have presented in this book. One might expect the statistical functionality of Python to grow over time.

If your sole objective is to do statistics, R is more attractive. Yet there are several reasons why you might prefer Python. You may already know Python and use it for other tasks. Indeed, it would be unusual for someone to solely do statistics. The data in this text is already clean and ready to use. In practice, this is rarely the case, and flexible software for obtaining and manipulating data is essential. You may already be using Python for this purpose.

Python also has a place at the heart of Machine Learning (ML), but this is a book about statistics rather than ML. But the aims of these two disciplines overlap considerably to the extent that any data analyst should become familiar with the ideas and methods of both. The datasets in this text are small by ML standards. I hope that a reader coming to this book from an ML background would learn new statistical perspectives on learning from data.

This book would not have been possible without several key open source Python packages. I thank the authors and maintainers of these packages for their outstanding work.

Chapter 1

Introduction

1.1 Before You Start

Statistics starts with a problem, proceeds with the collection of data, continues with the data analysis and finishes with conclusions. It is a common mistake of inexperienced statisticians to plunge into a complex analysis without paying attention to the objectives or even whether the data are appropriate for the proposed analysis. As Einstein said, the formulation of a problem is often more essential than its solution which may be merely a matter of mathematical or experimental skill.

To formulate the problem correctly, you must:

1. Understand the physical background. Statisticians often work in collaboration with others and need to understand something about the subject area. Regard this as an opportunity to learn something new rather than a chore.

2. Understand the objective. Again, often you will be working with a collaborator who may not be clear about what the objectives are. Beware of "fishing expeditions" — if you look hard enough, you will almost always find something, but that something may just be a coincidence.

3. Make sure you know what the client wants. You can often do quite different analyses on the same dataset. Sometimes statisticians perform an analysis far more complicated than the client really needed. You may find that simple descriptive statistics are all that are needed.

4. Put the problem into statistical terms. This is a challenging step and where irreparable errors are sometimes made. Once the problem is translated into the language of statistics, the solution is often routine. This is where human intelligence is decidedly superior to artificial intelligence. Defining the problem is hard to program. That a statistical method can read in and process the data is not enough. The results of an inapt analysis may be meaningless.

It is important to understand how the data were collected.

1. Are the data observational or experimental? Are the data a sample of convenience or were they obtained via a designed sample survey? How the data were collected has a crucial impact on what conclusions can be made.

2. Is there nonresponse? The data you do not see may be just as important as the data you do see.

3. Are there missing values? This is a common problem that is troublesome and time consuming to handle.

4. How are the data coded? In particular, how are the categorical variables represented?

5. What are the units of measurement?

6. Beware of data entry errors and other corruption of the data. This problem is all too common — almost a certainty in any real dataset of at least moderate size. Perform some data sanity checks.

1.2   Initial Data Analysis

This is a critical step that should always be performed. It is simple but it is vital. You should make numerical summaries such as means, standard deviations (SDs), maximum and minimum, correlations and whatever else is appropriate to the specific dataset. Equally important are graphical summaries. There is a wide variety of techniques to choose from. For one variable at a time, you can make boxplots, histograms, density plots and more. For two variables, scatterplots are standard while for even more variables, there are numerous good ideas for display including interactive and dynamic graphics. In the plots, look for outliers, data-entry errors, skewed or unusual distributions and structure. Check whether the data are distributed according to prior expectations.

Getting data into a form suitable for analysis by cleaning out mistakes and aberrations is often time consuming. It often takes more time than the data analysis itself. One might consider this the core work of *data science*. In this book, all the data will be ready to analyze, but you should realize that in practice this is rarely the case.

Let's look at an example. The National Institute of Diabetes and Digestive and Kidney Diseases conducted a study on 768 adult female Pima Indians living near Phoenix. The following variables were recorded: number of times pregnant, plasma glucose concentration at 2 hours in an oral glucose tolerance test, diastolic blood pressure (mmHg), triceps skin fold thickness (mm), 2-hour serum insulin (mu U/ml), body mass index (weight in kg/(height in m^2)), diabetes pedigree function, age (years) and a test whether the patient showed signs of diabetes (coded zero if negative, one if positive). The data may be obtained from UCI Repository of machine learning databases at archive.ics.uci.edu/ml.

Base Python has only limited functionality for numerical work. You will surely need to import some packages before you can accomplish anything. It is common to load all the packages you will need in a session at the beginning. We start with:

```
import pandas as pd
import numpy as np
import matplotlib.pyplot as plt
import scipy as sp
import seaborn as sns
import statsmodels.formula.api as smf
```

You can wait until you need them but it can be helpful when you share or return to your work later to have them all listed at the beginning so all will know which packages you need. The as pd means we can refer to functions in the pandas with the abbreviation pd.

Before doing anything else, one should find out the purpose of the study and more about how the data were collected. However, let's skip ahead to a look at the data:

```
import faraway.datasets.pima
pima = faraway.datasets.pima.load()
pima.head()
   pregnant  glucose  diastolic  triceps  insulin   bmi  diabetes  age  test
0         6      148         72       35        0  33.6     0.627   50     1
1         1       85         66       29        0  26.6     0.351   31     0
2         8      183         64        0        0  23.3     0.672   32     1
3         1       89         66       23       94  28.1     0.167   21     0
4         0      137         40       35      168  43.1     2.288   33     1
```

Many of the datasets used in this book are supplied in the faraway package. See the appendix for how to install this package. Any time you want to use one of these datasets, you will need to import the package containing the data you require and then load it.

The command pima.head() prints out the first five lines of the data frame. This is a good way to see what variables we have and what sort of values they take. You can type pima to see the whole data frame but 768 lines may be more than you want to examine.

If you want more details about the dataset, you can use:

```
print(faraway.datasets.pima.DESCR)
```

We start with some numerical summaries:

```
pima.describe().round(1)
       pregnant  glucose  diastolic  triceps  insulin    bmi  diabetes    age
count     768.0    768.0      768.0    768.0    768.0  768.0     768.0  768.0
mean        3.8    120.9       69.1     20.5     79.8   32.0       0.5   33.2
std         3.4     32.0       19.4     16.0    115.2    7.9       0.3   11.8
min         0.0      0.0        0.0      0.0      0.0    0.0       0.1   21.0
25%         1.0     99.0       62.0      0.0      0.0   27.3       0.2   24.0
50%         3.0    117.0       72.0     23.0     30.5   32.0       0.4   29.0
75%         6.0    140.2       80.0     32.0    127.2   36.6       0.6   41.0
max        17.0    199.0      122.0     99.0    846.0   67.1       2.4   81.0

        test
count  768.0
mean     0.3
std      0.5
min      0.0
25%      0.0
50%      0.0
75%      1.0
max      1.0
```

The describe() command is a quick way to get the usual univariate summary information. We round to one decimal place for compact, easier to read output. At this stage, we are looking for anything unusual or unexpected, perhaps indicating a data-entry error. For this purpose, a close look at the minimum and maximum values of each variable is worthwhile. Starting with pregnant, we see a maximum value of 17. This is large, but not impossible. However, we then see that the next five variables have minimum values of zero. No blood pressure is not good for the health — something must be wrong. Let's look at the first few sorted values:

```
pima['diastolic'].sort_values().head()
347    0
```

```
494    0
222    0
81     0
78     0
```

We see that at least the first 5 values are zero. We can count the zeroes:
```
np.sum(pima['diastolic'] == 0)
35
```

For one reason or another, the researchers did not obtain the blood pressures of 35 patients. In a real investigation, one would likely be able to question the researchers about what really happened. Nevertheless, this does illustrate the kind of misunderstanding that can easily occur. A careless statistician might overlook these presumed missing values and complete an analysis assuming that these were real observed zeroes. If the error was later discovered, they might then blame the researchers for using zero as a missing value code (not a good choice since it is a valid value for some of the variables) and not mentioning it in their data description. Unfortunately such oversights are not uncommon, particularly with datasets of any size or complexity. The statistician bears some share of responsibility for spotting these mistakes.

We set all zero values of the five variables to NaN which is a missing value code used by Python.
```
pima.replace({'diastolic' : 0, 'triceps' : 0, 'insulin' : 0,
    'glucose' : 0, 'bmi' : 0}, np.nan, inplace=True)
```
The variable test is not quantitative but categorical. Such variables are sometimes also called *factors*. However, because of the numerical coding, this variable has been treated as if it were quantitative. It is best to designate such variables as categorical so that they are treated appropriately. Sometimes people forget this and compute stupid statistics such as the "average zip code."
```
pima['test'] = pima['test'].astype('category')
pima['test'] = pima['test'].cat.rename_categories(
    ['Negative','Positive'])
pima['test'].value_counts()
Negative    500
Positive    268
```

Now that we have cleared up the missing values and coded the data appropriately, we are ready to do some plots. Perhaps the most well-known univariate plot is the histogram:
```
sns.distplot(pima.diastolic.dropna())
```
as seen in the first panel of Figure 1.1. We see a bell-shaped distribution for the diastolic blood pressures centered around 70. The construction of a histogram requires the specification of the number of bins and their spacing on the horizontal axis. Some choices can lead to histograms that obscure some features of the data. The seaborn package chooses the number and spacing of bins given the size and distribution of the data, but this choice is not foolproof and misleading histograms are possible. Some experimentation with other choices is sometimes worthwhile. Histograms are rough and some prefer to use kernel density estimates, which are essentially a smoothed version of the histogram (see Simonoff (1996) for a discussion of the relative merits of histograms and kernel estimates). This estimate is shown in the plot as a smooth curve.

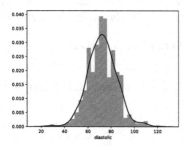 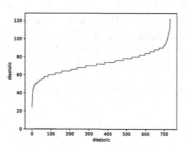

Figure 1.1   The first panel shows a histogram of the diastolic blood pressures, with a kernel density estimate superimposed. The second panel shows an index plot of the sorted values.

A simple alternative is to plot the sorted data against its index:
```
pimad = pima.diastolic.dropna().sort_values()
sns.lineplot(range(0, len(pimad)), pimad)
```
The advantage of this is that we can see all the cases individually. We can see the distribution and possible outliers. We can also see the discreteness in the measurement of blood pressure — values are rounded to the nearest even number and hence we see the "steps" in the plot.

Now we show a couple of bivariate plots, as seen in Figure 1.2:
```
sns.scatterplot(x='diastolic',y='diabetes',data=pima, s=20)
```
and
```
sns.boxplot(x="test", y="diabetes", data=pima)
```

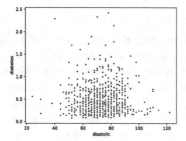 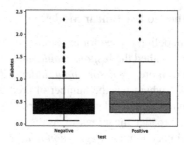

Figure 1.2   The first panel shows a scatterplot of the diastolic blood pressures against diabetes function and the second shows boxplots of diabetes function broken down by test result.

First, we see the standard scatterplot showing two quantitative variables. Second, we see a side-by-side boxplot suitable for showing a quantitative with a qualitative variable.

Sometimes we need to introduce a third variable into a bivariate plot. We show two different ways that the varying `test` result can be shown in the relationship between `diastolic` and `diabetes`:

```
sns.scatterplot(x="diastolic", y="diabetes", data=pima,
    style="test", alpha=0.3)
```

and

```
sns.relplot(x="diastolic", y="diabetes", data=pima, col="test")
```

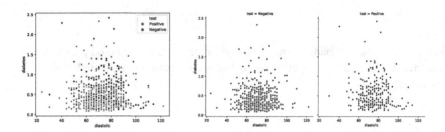

Figure 1.3   Two ways of distinguishing a factor variable in a bivariate scatterplot.

The first plot, shown in Figure 1.3, introduces the third element using the shape of the plotted point. The second plot uses two panels. Sometimes this is the better option when crowded plots make different colors or shapes hard to distinguish.

Good graphics are vital in data analysis. They help you avoid mistakes and suggest the form of the modeling to come. They are also important in communicating your analysis to others. Many in your audience or readership will focus on the graphs. This is your best opportunity to get your message over clearly and without misunderstanding. In some cases, the graphics can be so convincing that the formal analysis becomes just a confirmation of what has already been seen.

1.3   When to Use Linear Modeling

Linear modeling is used for explaining or modeling the relationship between a single variable Y, called the *response, outcome, output, endogenous* or *dependent* variable; and one or more *predictor, input, independent, exogenous* or *explanatory* variables, $X_1, \ldots, X_p$, where p is the number of predictors. We recommend you avoid using the words *independent* and *dependent* variables for X and Y, as these are easily confused with the broader meanings of terms. The endogenous/exogenous naming pair is popular in economics. *Regression analysis* is another term used for linear modeling although regressions can also be nonlinear.

When $p = 1$, it is called *simple* regression but when $p > 1$ it is called *multiple* regression or sometimes *multivariate* regression. When there is more than one response, then it is called *multivariate multiple* regression or sometimes (confusingly) multivariate regression. We will not cover this in this book, although you can just do separate regressions on each Y.

The response should be a continuous variable but if we are pedantic, all variables are measured with limited precision in practice and are therefore discrete.

Fortunately, provided the response variable is not measured too coarsely, we can ignore this objection without much consequence. The explanatory variables can be continuous, discrete or categorical, although we leave the handling of categorical explanatory variables to later in the book. Taking the example presented above, a regression with diastolic and bmi as Xs and diabetes as Y would be a multiple regression involving only quantitative variables which we tackle first. A regression with diastolic and test as Xs and bmi as Y would have one predictor that is quantitative and one that is qualitative, which we will consider later in Chapter 14 on *analysis of covariance*. A regression with test as X and diastolic as Y involves just qualitative predictors — a topic called *analysis of variance (ANOVA)*, although this would just be a simple two-sample situation. A regression of test as Y on diastolic and bmi as predictors would involve a qualitative response. A *logistic regression* could be used, but this will not be covered in this book.

Regression analyses have two main objectives:

1. Prediction of future or unseen responses given specified values of the predictors.

2. Assessment of the effect of, or relationship between, explanatory variables and the response. We would like to infer causal relationships if possible.

You should be clear on the objective for the given data because some aspects of the resulting analysis may differ. Regression modeling can also be used in a descriptive manner to summarize the relationships between the variables. However, most end users of data have more specific questions in mind and want to direct the analysis toward a particular set of goals.

It is rare, except in a few cases in the precise physical sciences, to know (or even suspect) the true model. In most applications, the model is an empirical construct designed to answer questions about prediction or causation. It is usually not helpful to think of regression analysis as the search for some true model. The model is a means to an end, not an end in itself.

1.4 History

In the 18th century, accurate navigation was a difficult problem of commercial and military interest. Although it is relatively easy to determine latitude from Polaris, also known as the North Star, finding longitude then was difficult. Various attempts were made to devise a method using astronomy. Contrary to popular supposition, the moon does not always show the same face and moves such that about 60% of its surface is visible at some time.

Tobias Mayer collected data on the locations of various landmarks on the moon, including the Manilius crater, as they moved relative to the earth. He derived an equation describing the motion of the moon (called *libration*) taking the form:

$$arc = \alpha + \beta sinang + \gamma cosang$$

He wished to obtain values for the three unknowns α, β and γ. The variables arc, sinang and cosang can be observed using a telescope. A full explanation of the story behind the data and the derivation of the equation can be found in Stigler (1986).

Since there are three unknowns, we need only three distinct observations of the set of three variables to find a unique solution for α, β and γ. Embarassingly for Mayer, there were 27 sets of observations available. Astronomical measurements were naturally subject to some variation and so there was no solution that fit all 27 observations. Let's take a look at the first few lines of the data:

```
import faraway.datasets.manilius
manilius = faraway.datasets.manilius.load()
manilius.head()
        arc  sinang  cosang  group
0  13.166667  0.8836  -0.4682      1
1  13.133333  0.9996  -0.0282      1
2  13.200000  0.9899   0.1421      1
3  14.250000  0.2221   0.9750      3
4  14.700000  0.0006   1.0000      3
```

Mayer's solution was to divide the data into three groups so that observations within each group were similar in some respect. He then computed the sum of the variables within each group. We can also do this:

```
moon3 = manilius.groupby('group').sum()
moon3
              arc  sinang   cosang
group
1      118.133333  8.4987  -0.7932
2      140.283333 -6.1404   1.7443
3      127.533333  2.9777   7.9649
```

Now there are just three equations in three unknowns to be solved. The solution is:

```
moon3['intercept'] = [9]*3
np.linalg.solve(moon3[['intercept','sinang','cosang']],
    moon3['arc'])
array([14.54458591, -1.48982207,  0.13412639])
```

Hence the computed values of α, β and γ are 14.5, -1.49 and 0.134, respectively. One might question how Mayer selected his three groups, but this solution does not seem unreasonable.

Similar problems with more linear equations than unknowns continued to arise until 1805, when Adrien Marie Legendre published the method of least squares. Suppose we recognize that the equation is not exact and introduce an error term, ε:

$$\text{arc}_i = \alpha + \beta \sin\text{ang}_i + \gamma \cos\text{ang}_i + \varepsilon_i$$

where $i = 1, \ldots, 27$. Now we find α, β and γ that minimize the sum of the squared errors: $\sum \varepsilon^2$. We will investigate this in much greater detail in the chapter to follow, but for now we simply present the solution using the smf.ols function from statsmodels:

```
mod = smf.ols('arc ~ sinang + cosang', manilius).fit()
mod.params
Intercept    14.561624
sinang       -1.504581
cosang        0.091365
```

We observe that this solution is quite similar to Mayer's. The least squares solution is more satisfactory in that it requires no arbitrary division into groups. Carl Friedrich Gauss claimed to have devised the method of least squares earlier but

without publishing it. At any rate, he did publish in 1809 showing that the method
of least squares was, in some sense, optimal.

For many years, the method of least squares was confined to the physical sciences
where it was used to resolve problems of overdetermined linear equations. The equations were derived from theory, and least squares was used as a method to fit data to
these equations to estimate coefficients like α, β and γ above. It was not until later in
the 19th century that linear equations (or models) were suggested empirically from
the data rather than from theories of physical science. This opened up the field to the
social and life sciences.

Francis Galton, a nephew of Charles Darwin, was important in this extension of
statistics into social science. He coined the term *regression to mediocrity* in 1875
from which the rather peculiar term *regression* derives. Let's see how this terminology arose by looking at one of the datasets he collected at the time on the heights of
parents and children in Galton (1886). We load and plot the data as seen in Figure 1.4.

```
import faraway.datasets.families
families = faraway.datasets.families.load()
sns.scatterplot(x='midparentHeight', y='childHeight',
    data=families, s=20)
```

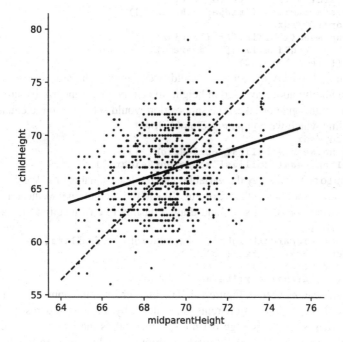

Figure 1.4   The height of a child is plotted against a combined parental height defined as (father's height + 1.08 × mother's height)/2.

We see that `midparentHeight`, defined as the father's height plus 1.08 times the
mother's height divided by two, is correlated with the `childHeight`, both in inches.

Now we might propose a linear relationship between the two of the form:

$$\texttt{childHeight} = \alpha + \beta \texttt{midparentHeight} + \varepsilon$$

We can estimate α and β using smf.ols().

```
mod = smf.ols('childHeight ~ midparentHeight', families).fit()
mod.params
Intercept          22.636241
midparentHeight     0.637361
```

For the simple case of a response y and a single predictor x, we can write the equation in the form:

$$\frac{y-\bar{y}}{SD_y} = r\frac{(x-\bar{x})}{SD_x}$$

where r is the correlation between x and y. The equation can be expressed in words as: the response in standard units is the correlation times the predictor in standard units. We can verify that this produces the same results as above by rearranging the equation in the form $y = \alpha + \beta x$ and computing the estimates:

```
cor = sp.stats.pearsonr(families['childHeight'],
    families['midparentHeight'])[0]
sdy = np.std(families['childHeight'])
sdx = np.std(families['midparentHeight'])
beta = cor*sdy/sdx
alpha = np.mean(families['childHeight']) - \
    beta*np.mean(families['midparentHeight'])
np.round([alpha,beta],2)
```

Now one might naively expect that a child with parents who are, for example, one standard deviation above average in height, to also be one standard deviation above average in height, give or take. The supposition would set $r = 1$ in the equation and leads to a line which we compute and plot below:

```
beta1 = sdy/sdx
alpha1 = np.mean(families['childHeight']) - \
    beta1*np.mean(families['midparentHeight'])
```

We use lmplot() to display the variables and the least squares line. We do not want the confidence band, hence the ci=None. To add the second line, we need to specify the range in the horizontal scale and draw a dashed line connecting the calculated points at each end.

```
sns.lmplot('midparentHeight', 'childHeight', families,
    ci=None, scatter_kws={'s':2})
xr = np.array([64,76])
plt.plot(xr, alpha1 + xr*beta1,'--')
```

The result can be seen in Figure 1.4. The lines cross at the point of the averages. We can see that a child of tall parents is predicted by the least squares line to have a height which is above average but not quite as tall as the parents, as the dashed line would have you believe. Similarly children of below average height parents are predicted to have a height which is still below-average but not quite as short as the parents. This is why Galton used the phrase "regression to mediocrity" and the phenomenom is sometimes called the regression effect.

This applies to any (x,y) situation like this. For example, in sports, an athlete may have a spectacular first season only to do not quite as well in the second season.

Sports writers come up with all kinds of explanations for this but the regression effect is likely to be the unexciting cause. In the parents and children example, although it does predict that successive descendants in the family will come closer to the mean, it does not imply the same of the population in general since random fluctuations will maintain the variation, so no need to get too pessimistic about mediocrity! In many other applications of linear modeling, the regression effect is not of interest because different types of variables are measured. Unfortunately, we are now stuck with the rather gloomy word of regression thanks to Galton.

Regression methodology developed rapidly with the advent of high-speed computing. Just fitting a regression model used to require extensive hand calculation. As computing hardware has improved, the scope for analysis has widened. This has led to an extensive development in the methodology and the scale of problems that can be tackled.

Exercises

Not all the answers to the questions below can be derived from code illustrated in this chapter. You may need to resort to internet Python resources.

1. The dataset `teengamb` concerns a study of teenage gambling in Britain.

   (a) Turn the sex variable into a categorical variable with appropriate labels. Count the number in each category.

   (b) Use both the `boxplot` and the `swarmplot` functions from `seaborn` to plot the status broken down by sex. Contrast the two plotting methods.

   (c) Use both the `distplot` and the `countplot` functions from `seaborn` to show the distributions of the verbal scores. Do not show the smoothed density on the `distplot`. Contrast the two plotting methods - which is best here?

   (d) Plot the gamble as the response and income as the predictor broken down by sex. Make two plots, one with a single frame where sex is distinguished by the color of the point and another where the sexes appear in different frames. Which plot do you prefer and why?

   (e) Construct a summary statistics table of numerical variables. Can you tell which variable is highly skewed from the table?

2. The dataset `uswages` is drawn as a sample from the Current Population Survey in 1988.

   (a) Construct a subset of the data with only the wage and four geographical variables.

   (b) A weighted mean is given by $\sum w_i y_i / \sum w_i$ for weights w and data y. Compute the mean wage in the north-east using this formula.

   (c) Compute the mean wage in the north-east using the `groupby` function from `pandas`. This should also give you the mean wage for those not living in the north-east.

   (d) Compute the row sums for just the geographic variables. What value do they take?

(e) The subset matrix of geographic variables can be called a dummy matrix where ones and zeroes are used to code a categorical variable. Reconstruct an area categorical variable which takes the four possible values.

(f) Make a boxplot of the wage broken down by area.

(g) Repeat the previous plot but on a log scale. Which is preferable?

3. The dataset `prostate` is from a study on 97 men with prostate cancer who were due to receive a radical prostatectomy.

(a) Use the `pairplot` function from `seaborn` to construct an array of scatterplots of the first four variables.

(b) Compute the correlations of the first four variables.

(c) The `lbph` variable is on the log scale. Many cases take the minimum value of this variable. What value of benign prostatic hyperplasia do you think this represents?

(d) Use the `distplot` function from `seaborn` to make a histogram of the ages using the rug option. Create a version where the bin width is one year. Contrast the two plots.

(e) Use the `melt` function from `pandas` with `lpsa` as the id variable to produce a long version of the dataset. Now use `replot` to produce a grid of 8 scatterplots of the data where `lpsa` is the response.

4. The dataset `sat` comes from a study entitled "Getting What You Pay For: The Debate Over Equity in Public School Expenditures."

(a) Verify that the sum of the verbal and math scores equals the total score.

(b) Compare the distributions of verbal and math scores using `jointplot` from `seaborn`. Are they similar?

(c) Standardize both the verbal and math scores. Plot the standardized scores with verbal on the x-axis. Plot the $y = x$ line.

(d) Fit a linear model with math as the response and verbal as the predictor. Show the estimated slope and compare it with the correlation between these two variables. Comment.

(e) Fit another linear model with the roles of the predictor and response exchanged. Why is the estimated slope the same? Is the fitted line from this model and the previous model the same?

(f) Make predictions for the following students. (i) Predict the math score of a student scoring 2SDs above average on the verbal test. (ii) Predict the verbal score of a student scoring 2SDs above average on the math test. (iii) Predict the math score of a student with an average score on the verbal test. (iv) Predict the math score of a student with no information about their verbal score.

5. The dataset `divusa` contains data on divorces in the United States from 1920 to 1996.

(a) Make a plot each with `lineplot` and `scatterplot` from `seaborn`. Put the year on the x-axis and the divorce rate on the y-axis. Compare the two plots.

(b) Use the shift function from pandas to plot divorce rate from the current year against the divorce rate for the previous year. Does this show that one could reasonably predict the divorce rate for the following year by using the divorce rate from the current year?

(c) Fit a linear model with the divorce rate as the response and the year as the predictor. In what year does the model predict the divorce rate to hit 100%? Is this a reasonable prediction?

(d) Use the scatterplot function from seaborn to make a plot with femlab on the x-axis, divorce rate on the y-axis and the color of the point changing with the year.

Chapter 2

Estimation

2.1 Linear Model

Let's start by defining what is meant by a linear model. Suppose we want to model the response Y in terms of three predictors, X_1, X_2 and X_3. One very general form for the model would be:

$$Y = f(X_1, X_2, X_3) + \varepsilon$$

where f is some unknown function and ε is the error in this representation. ε is additive in this instance, but could enter in some even more general form. Still, if we assume that f is a smooth, continuous function, that still leaves a very wide range of possibilities. Even with just three predictors, we would need a substantial amount of data to try to estimate f directly. With smaller datasets, we usually have to assume that it has some more restricted form, perhaps linear as in:

$$Y = \beta_0 + \beta_1 X_1 + \beta_2 X_2 + \beta_3 X_3 + \varepsilon$$

where β_i, $i = 0, 1, 2, 3$ are unknown *parameters*. Unfortunately this term is subject to some confusion as engineers often use the term *parameter* for what statisticians call the variables, Y, X_1 and so on. β_0 is called the *intercept* term.

Thus the problem is reduced to the estimation of four parameters rather than the infinite dimensional f. In a linear model the *parameters enter linearly* — the predictors themselves do not have to be linear. For example:

$$Y = \beta_0 + \beta_1 X_1 + \beta_2 \log X_2 + \beta_3 X_1 X_2 + \varepsilon$$

is a linear model, but:

$$Y = \beta_0 + \beta_1 X_1^{\beta_2} + \varepsilon$$

is not. Some relationships can be transformed to linearity — for example, $y = \beta_0 x_1^{\beta} \varepsilon$ can be linearized by taking logs. Linear models seem rather restrictive, but because the predictors can be transformed and combined in any way, they are actually very flexible. The term *linear* is often used in everyday speech as almost a synonym for simplicity. This gives the casual observer the impression that linear models can only handle small, simple datasets. This is far from the truth — linear models can easily be expanded and modified to handle complex datasets. *Linear* is also used to refer to straight lines, but linear models can be curved, by adding quadratic terms for example. Truly nonlinear models are rarely absolutely necessary and most often arise from a theory about the relationships between the variables, rather than an empirical investigation.

Where do models come from? We distinguish several different sources:

1. Physical theory may suggest a model. For example, Hooke's law says that the extension of a spring is proportional to the weight attached. Models like these usually arise in the physical sciences and engineering.

2. Experience with past data. Similar data used in the past were modeled in a particular way. It is natural to see whether the same model will work with the current data. Models like these usually arise in the social sciences.

3. No prior idea exists — the model comes from an exploration of the data. We use skill and judgment to pick a model. Sometimes it does not work and we have to try again.

Models that derive directly from physical theory are relatively uncommon so that usually the linear model can only be regarded as an approximation to a complex reality. We hope it predicts well or explains relationships usefully, but usually we do not believe it is exactly true. A good model is like a map that guides us to our destination. For the rest of this chapter, we will stay in the special world of Mathematics where all models are true.

2.2   Matrix Representation

We want a general solution to estimating the parameters of a linear model. We can find simple formulae for some special cases but to devise a method that will work in all cases, we need to use matrix algebra. Let's see how this can be done.

We start with some data where we have a response Y and, say, three predictors, X_1, X_2 and X_3. The data might be presented in tabular form like this:

$$
\begin{array}{cccc}
y_1 & x_{11} & x_{12} & x_{13} \\
y_2 & x_{21} & x_{22} & x_{23} \\
\cdots & & \cdots & \\
y_n & x_{n1} & x_{n2} & x_{n3}
\end{array}
$$

where n is the number of observations, or *cases*, in the dataset.

Given the actual data values, we may write the model as:

$$y_i = \beta_0 + \beta_1 x_{i1} + \beta_2 x_{i2} + \beta_3 x_{i3} + \varepsilon_i \quad i = 1, \ldots, n.$$

It will be more convenient to put this in a matrix/vector representation. The regression equation is then written as:

$$y = X\beta + \varepsilon$$

where $y = (y_1, \ldots, y_n)^T$, $\varepsilon = (\varepsilon_1, \ldots, \varepsilon_n)^T$, $\beta = (\beta_0, \ldots, \beta_3)^T$ and:

$$
X = \begin{pmatrix}
1 & x_{11} & x_{12} & x_{13} \\
1 & x_{21} & x_{22} & x_{23} \\
\cdots & & \cdots & \\
1 & x_{n1} & x_{n2} & x_{n3}
\end{pmatrix}
$$

The column of ones incorporates the intercept term. One simple example is the *null model* where there is no predictor and just a mean $y = \mu + \varepsilon$:

$$
\begin{pmatrix} y_1 \\ \ldots \\ y_n \end{pmatrix} = \begin{pmatrix} 1 \\ \ldots \\ 1 \end{pmatrix} \mu + \begin{pmatrix} \varepsilon_1 \\ \ldots \\ \varepsilon_n \end{pmatrix}
$$

We can assume that $E\varepsilon = 0$ since if this were not so, we could simply absorb the nonzero expectation for the error into the mean μ to get a zero expectation.

2.3   Estimating β

The regression model, $y = X\beta + \varepsilon$, partitions the response into a systematic component $X\beta$ and a random component ε. We would like to choose β so that the systematic part explains as much of the response as possible. Geometrically speaking, the response lies in an n-dimensional space, that is, $y \in \mathbb{R}^n$ while $\beta \in \mathbb{R}^p$ where p is the number of parameters. If we include the intercept, then p is the number of predictors plus one. It is easy to get confused as to whether p is the number of predictors or parameters, as different authors use different conventions, so be careful.

The problem is to find β so that $X\beta$ is as close to Y as possible. The best choice, the estimate $\hat{\beta}$, is apparent in the geometrical representation seen in Figure 2.1. $\hat{\beta}$ is, in this sense, the best estimate of β within the model space. The $\hat{\beta}$ values are sometimes called the regression coefficients. The response predicted by the model is $\hat{y} = X\hat{\beta}$ or Hy where H is an orthogonal projection matrix. The $\hat{y}$ are called *predicted* or *fitted* values. The difference between the actual response and the predicted response is denoted by $\hat{\varepsilon}$ and is called the *residual*.

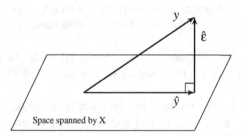

Figure 2.1   Geometrical representation of the estimation β. The data vector Y is projected orthogonally onto the model space spanned by X. The fit is represented by projection $\hat{y} = X\hat{\beta}$ with the difference between the fit and the data represented by the residual vector $\hat{\varepsilon}$.

The conceptual purpose of the model is to represent, as accurately as possible, something complex, y, which is n-dimensional, in terms of something much simpler, the model, which is p-dimensional. Thus if our model is successful, the structure in the data should be captured in those p dimensions, leaving just random variation in the residuals which lie in an $(n - p)$-dimensional space. We have:

$$\text{Data} = \quad \text{Systematic Structure} \quad + \quad \text{Random Variation}$$
$$n \text{ dimensions} = \quad p \text{ dimensions} \quad + \quad (n-p) \text{ dimensions}$$

2.4 Least Squares Estimation

The estimation of β can also be considered from a nongeometric point of view. We define the best estimate of β as the one which minimizes the sum of the squared errors:

$$\sum \varepsilon_i^2 = \varepsilon^T \varepsilon = (y - X\beta)^T (y - X\beta)$$

Differentiating with respect to β and setting to zero, we find that $\hat{\beta}$ satisfies:

$$X^T X \hat{\beta} = X^T y$$

These are called the *normal equations*. We can derive the same result using the geometric approach. Now provided $X^T X$ is invertible:

$$
\begin{aligned}
\hat{\beta} &= (X^T X)^{-1} X^T y \\
X\hat{\beta} &= X(X^T X)^{-1} X^T y \\
\hat{y} &= Hy
\end{aligned}
$$

$H = X(X^T X)^{-1} X^T$ is called the *hat matrix* and is the orthogonal projection of y onto the space spanned by X. H is useful for theoretical manipulations, but you usually do not want to compute it explicitly, as it is an $n \times n$ matrix which could be uncomfortably large for some datasets. The following useful quantities can now be represented using H.

The predicted or fitted values are $\hat{y} = Hy = X\hat{\beta}$ while the residuals are $\hat{\varepsilon} = y - X\hat{\beta} = y - \hat{y} = (I - H)y$. The residual sum of squares (RSS) is $\hat{\varepsilon}^T \hat{\varepsilon} = y^T (I - H)^T (I - H)y = y^T (I - H)y$.

Later, we will show that the least squares estimate is the best possible estimate of β when the errors ε are uncorrelated and have equal variance. We can express this fact more compactly as var $\varepsilon = \sigma^2 I$ meaning that the covariance matrix of ε is a diagonal matrix with values of σ^2 along that diagonal. $\hat{\beta}$ is unbiased and has variance $(X^T X)^{-1} \sigma^2$ provided var $\varepsilon = \sigma^2 I$. Since $\hat{\beta}$ is a vector, its variance is a matrix.

We also need to estimate σ^2. We find that $E\hat{\varepsilon}^T \hat{\varepsilon} = \sigma^2 (n - p)$, which suggests the estimator:

$$\hat{\sigma}^2 = \frac{\hat{\varepsilon}^T \hat{\varepsilon}}{n - p} = \frac{\text{RSS}}{n - p}$$

as an unbiased estimate of σ^2. $n - p$ is called the *degrees of freedom* of the model. Sometimes you need the standard error for a particular component of $\hat{\beta}$ which can be picked out as $se(\hat{\beta}_{i-1}) = \sqrt{(X^T X)^{-1}_{ii}} \hat{\sigma}$.

2.5 Examples of Calculating $\hat{\beta}$

In a few simple models, it is possible to derive explicit formulae for $\hat{\beta}$:

1. When $y = \mu + \varepsilon$, $X = \mathbf{1}$ and $\beta = \mu$ hence $X^T X = \mathbf{1}^T \mathbf{1} = n$ so:

$$\hat{\beta} = (X^T X)^{-1} X^T y = \frac{1}{n} \mathbf{1}^T y = \bar{y}$$

2. Simple linear regression (one predictor):

$$y_i = \beta_0 + \beta_1 x_i + \varepsilon_i$$

$$\begin{pmatrix} y_1 \\ \dots \\ y_n \end{pmatrix} = \begin{pmatrix} 1 & x_1 \\ & \dots \\ 1 & x_n \end{pmatrix} \begin{pmatrix} \beta_0 \\ \beta_1 \end{pmatrix} + \begin{pmatrix} \varepsilon_1 \\ \dots \\ \varepsilon_n \end{pmatrix}$$

We can now apply the formula but a simpler approach is to rewrite the equation as:

$$y_i = \overbrace{\beta_0 + \beta_1 \bar{x}}^{\beta_0'} + \beta_1 (x_i - \bar{x}) + \varepsilon_i$$

so now:

$$X = \begin{pmatrix} 1 & x_1 - \bar{x} \\ & \dots \\ 1 & x_n - \bar{x} \end{pmatrix} \qquad X^T X = \begin{pmatrix} n & 0 \\ 0 & \sum_{i=1}^n (x_i - \bar{x})^2 \end{pmatrix}$$

Next work through the rest of the calculation to reconstruct the familiar estimate, that is:

$$\hat{\beta}_1 = \frac{\sum (x_i - \bar{x}) y_i}{\sum (x_i - \bar{x})^2}$$

In higher dimensions, it is usually not possible to find such explicit formulae for the parameter estimates unless $X^T X$ happens to be a simple form. So typically we need computers to fit such models. Regression has a long history, so in the time before computers became readily available, fitting even quite simple models was a tedious time-consuming task. When computing was expensive, data analysis was limited. It was designed to keep calculations to a minimum and restrict the number of plots. This mindset remained in statistical practice for some time even after computing became widely and cheaply available. Now it is a simple matter to fit a multitude of models and make more plots than one could reasonably study. The challenge now for the analyst is to choose among these intelligently to extract the crucial information in the data.

2.6 Example

Now let's look at an example concerning the number of species found on the various Galápagos Islands. As before, we start by loading the packages we will need in this chapter:

```
import pandas as pd
import numpy as np
import matplotlib.pyplot as plt
import scipy as sp
import statsmodels.api as sm
import statsmodels.formula.api as smf
import seaborn as sns
```

There are 30 cases (Islands) and seven variables in the dataset. We start by reading
the data into Python and examining it.

```
import faraway.datasets.galapagos
galapagos = faraway.datasets.galapagos.load()
galapagos.head()
```

	Species	Area	Elevation	Nearest	Scruz	Adjacent
Baltra	58	25.09	346	0.6	0.6	1.84
Bartolome	31	1.24	109	0.6	26.3	572.33
Caldwell	3	0.21	114	2.8	58.7	0.78
Champion	25	0.10	46	1.9	47.4	0.18
Coamano	2	0.05	77	1.9	1.9	903.82

The variables are Species — the number of species found on the island, Area —
the area of the island (km^2), Elevation — the highest elevation of the island (m),
Nearest — the distance from the nearest island (km), Scruz — the distance from
Santa Cruz Island (km), Adjacent — the area of the adjacent island (km^2).

The data were presented by Johnson and Raven (1973) and also appear in Weisberg (1985). I have filled in some missing values for simplicity (see Chapter 13
for how this can be done). Fitting a linear model in Python can be done using
the smf.ols() command. Notice the syntax for specifying the predictors in the
model. This is part of the *Wilkinson–Rogers* notation. It is convenient that the data,
galapagos, is a Pandas DataFrame. We have explicitly specified the formula and
data arguments here although this is not necessary and we will generally skip doing
this in the future.

```
lmod = smf.ols(
    formula='Species ~ Area+Elevation+Nearest+Scruz+Adjacent',
    data=galapagos).fit()
lmod.summary()
```

                          OLS Regression Results

==
Dep. Variable:	Species	R-squared:	0.766
Model:	OLS	Adj. R-squared:	0.717
Method:	Least Squares	F-statistic:	15.70
Date:	Mon, 01 xxx 20xx	Prob (F-statistic):	6.84e-07
Time:	11:29:12	Log-Likelihood:	-162.54
No. Observations:	30	AIC:	337.1
Df Residuals:	24	BIC:	345.5
Df Model:	5		
Covariance Type:	nonrobust		
==

	coef	std err	t	P>|t|	[0.025	0.975]
Intercept	7.0682	19.154	0.369	0.715	-32.464	46.601
Area	-0.0239	0.022	-1.068	0.296	-0.070	0.022
Elevation	0.3195	0.054	5.953	0.000	0.209	0.430
Nearest	0.0091	1.054	0.009	0.993	-2.166	2.185
Scruz	-0.2405	0.215	-1.117	0.275	-0.685	0.204

EXAMPLE   21

| Adjacent | -0.0748 | 0.018 | -4.226 | 0.000 | -0.111 | -0.038 |

Omnibus:	12.683	Durbin-Watson:	2.476
Prob(Omnibus):	0.002	Jarque-Bera (JB):	13.498
Skew:	1.136	Prob(JB):	0.00117
Kurtosis:	5.374	Cond. No.	1.90e+03

Warnings:
[1] Standard Errors assume that the covariance matrix of the errors is correctly specified.
[2] The condition number is large, 1.9e+03. This might indicate that there are strong multicollinearity or other numerical problems.

For my tastes, this output contains rather too much information — certainly more than we want to consider now. I have written an alternative called sumary which produces a shorter version of this. Since we will be looking at a lot of regression output, the use of this version makes this book several pages shorter. Of course, if you prefer the above, feel free to add the extra "m" in the function call. You will need to install (see Appendix) and import my package if you want to use my version:

```
import faraway.utils
lmod.sumary()
          coefs stderr tvalues pvalues
Intercept 7.068 19.154   0.37  0.7154
Area      -0.024  0.022  -1.07  0.2963
Elevation  0.319  0.054   5.95  0.0000
Nearest    0.009  1.054   0.01  0.9932
Scruz     -0.241  0.215  -1.12  0.2752
Adjacent  -0.075  0.018  -4.23  0.0003
```

n=30 p=6 Residual SD=60.975 R-squared=0.77

We can identify several useful quantities in this output. Other statistical packages tend to produce output quite similar to this. One useful feature of Python is that it is possible to directly calculate quantities of interest. Of course, it is not necessary here because the smf.ols() function does the job, but it is very useful when the statistic you want is not part of the prepackaged functions. First, we create the X-matrix, by first taking all but the first column in the DataFrame and then inserting a column of ones:

```
X = galapagos.iloc[:,1:]
X.insert(0,'intercept',1)
```

Now let's construct $(X^T X)^{-1}$. The .T does transpose and @ does matrix multiplication. np.linalg.inv(A) computes A^{-1}

```
XtXi = np.linalg.inv(X.T @ X)
```

We can get $\hat{\beta}$ directly, using $(X^T X)^{-1} X^T y$:

```
(XtXi @ X.T) @ galapagos.Species
0    7.068221
1   -0.023938
2    0.319465
3    0.009144
4   -0.240524
5   -0.074805
```

This is a very bad way to compute $\hat{\beta}$. It is inefficient because it is expensive to compute a matrix inverse and we can avoid that here. It can be very inaccurate when the predictors are strongly correlated. Such problems are exacerbated by large datasets. A better, but not perfect, way is:

```
np.linalg.solve(X.T @ X, X.T @ galapagos.Species)
array([ 7.06822071, -0.02393834,  0.31946476,  0.00914396, -0.24052423,
       -0.07480483])
```

where `np.linalg.solve(A,b)` solves $Ax = b$. Here we get the same result as `smf.ols()` because the data are well behaved. In the long run, you are advised to use carefully programmed code such as found in `statsmodels` or other packages. By default, `statsmodels` uses the Moore-Penrose inverse method. One can also use the QR decomposition. See Section 2.7 for more on this.

We can extract the regression quantities we need from the model object. Commonly used are `.resid`, `.fittedvalues` which give the residuals and the fitted values respectively. `.df_resid` which gives the degrees of freedom, `.ssr` which gives the RSS (sum of squares for the residuals) and `.params` which gives the $\hat{\beta}$. Further quantities are available — consult the help for `RegressionResults`.

We can estimate σ using the formula in the text above or extract it from the summary object:

```
np.sqrt(lmod.mse_resid)
60.975
```

2.7 Computing Least Squares Estimates

This section might be skipped unless you are interested in the actual calculation of $\hat{\beta}$ and related quantities. The Moore-Penrose inverse of an $n \times p$ matrix X is an $p \times n$ matrix X^- satisfying the following conditions:

(i) $XX^-X = X$

(ii) $X^-XX^- = X^-$

(iii) $(XX^-)^T = XX^-$

(iv) $(X^-X)^T = X^-X$

When X has full rank p, we have

$$X^- = (X^TX)^{-1}X^T$$

so that we may obtain the least squares estimate as $\hat{\beta} = X^-y$. Furthermore, we can show that even when X is not full rank, this solution minimizes the sum of squares. The particular advantage of this solution is that it minimizes the Euclidean norm of β, $||\beta||$. We can obtain the Moore-Penrose inverse using a singular value decomposition (SVD) with:

```
Xmp = np.linalg.pinv(X)
Xmp.shape
(6, 30)
```

We see this has the expected dimension. We compute the least squares estimates as

```
Xmp @ galapagos.Species
array([ 7.06822071, -0.02393834,  0.31946476,  0.00914396, -0.24052423,
       -0.07480483])
```

This is the default method used by `statsmodels` to compute least squares estimates.

The QR decomposition is an alternative to the Moore-Penrose inverse method for computing $\hat{\beta}$. It is the default choice of R. Any design matrix X can be written as:

$$X = Q \begin{pmatrix} R \\ 0 \end{pmatrix} = Q_f R$$

where Q is an $n \times n$ orthogonal matrix, that is $Q^T Q = QQ^T = I$ and R is a $p \times p$ upper triangular matrix ($R_{ij} = 0$ for $i > j$). The 0 is an $(n - p) \times p$ matrix of zeroes, while Q_f is the first p columns of Q.

The $RSS = (y - X\beta)^T (y - X\beta) = \|y - X\beta\|^2$ where $\| \cdot \|$ is the Euclidean length of a vector. The matrix Q represents a rotation and does not change length. Hence:

$$RSS = \|Q^T y - Q^T X\beta\|^2 = \left\| \begin{pmatrix} f \\ r \end{pmatrix} - \begin{pmatrix} R \\ 0 \end{pmatrix} \beta \right\|^2$$

where $\begin{pmatrix} f \\ r \end{pmatrix} = Q^T y$ for vector f of length p and vector r of length $n - p$. From this we see:

$$RSS = \|f - R\beta\|^2 + \|r\|^2$$

which can be minimized by setting β so that $R\beta = f$.

Let's see how this works for the Galápagos data. First we compute the QR decomposition:

```
q, r = np.linalg.qr(X)
```

We can compute f:

```
f = q.T @ galapagos.Species
f
array([-466.84219318,  381.40557435,  256.25047255,    5.40764552,
       -119.49834019,  257.69436853])
```

Solving $R\beta = f$ is computationally easy because of the triangular form of R. We use the method of backsubstitution:

```
sp.linalg.solve_triangular(r, f)
array([ 7.06822071, -0.02393834,  0.31946476,  0.00914396, -0.24052423,
       -0.07480483])
```

where the results match those seen previously.

We can ask statsmodels to use the QR method in preference to the default Moore-Penrose with:

```
lmodform = smf.ols(
    'Species ~ Area + Elevation + Nearest + Scruz  + Adjacent',
    galapagos)
lmod = lmodform.fit(method="qr")
lmod.params
Intercept    7.068221
Area        -0.023938
Elevation    0.319465
Nearest      0.009144
Scruz       -0.240524
Adjacent    -0.074805
```

The results are, as expected, the same.

For full rank X models, there will be very little difference between these two methods. The choice makes a difference when X is not full rank (or very close to it). The Moore-Penrose method will produce a solution in these circumstances whereas the QR method will require some modification. We discuss this further in Section 2.10.

An alternative within Python is to use the general least squares problem solver `lstsq` from the `scipy` package:

```
params, res, rnk, s = sp.linalg.lstsq(X, galapagos['Species'])
params
array([ 7.06822071, -0.02393834,  0.31946476,  0.00914396, -0.24052423,
       -0.07480483])
```

Again, this produces the same outcome. This method uses an SVD and will produce a solution in the less than full rank case. The popular `scikit-learn` machine learning package does regression based on this function. The important difference is that this package does not compute the full range of subsidiary information obtained with `statsmodels`. Since we will usually want this extra information in this book, we will mostly use `statsmodels`. However, if your application requires only $\hat{\beta}$ and the size of the dataset means that speed is a concern, you should choose accordingly.

2.8 Gauss–Markov Theorem

$\hat{\beta}$ is a plausible estimator, but there are alternatives. Nonetheless, there are three good reasons to use least squares:

1. It results from an orthogonal projection onto the model space. It makes sense geometrically.
2. If the errors are independent and identically normally distributed, it is the maximum likelihood estimator. Loosely put, the maximum likelihood estimate is the value of β that maximizes the probability of the data that was observed.
3. The Gauss–Markov theorem states that $\hat{\beta}$ is the best linear unbiased estimate (BLUE).

To understand the Gauss–Markov theorem we first need to understand the concept of an *estimable function*. A linear combination of the parameters $\psi = c^T \beta$ is estimable if and only if there exists a linear combination $a^T y$ such that:

$$Ea^T y = c^T \beta \qquad \forall \beta$$

Estimable functions include predictions of future observations, which explains why they are well worth considering. If X is of full rank, then all linear combinations are estimable.

Suppose $E\varepsilon = 0$ and var $\varepsilon = \sigma^2 I$. Suppose also that the structural part of the model, $EY = X\beta$ is correct. (Clearly these are big assumptions and so we will address the implications of this later.) Let $\psi = c^T \beta$ be an estimable function; then the Gauss–Markov theorem states that in the class of all unbiased linear estimates of ψ, $\hat{\psi} = c^T \hat{\beta}$ has the minimum variance and is unique.

We prove this theorem. Suppose $a^T y$ is some unbiased estimate of $c^T \beta$ so that:

$$\begin{aligned} Ea^T y &= c^T \beta \qquad \forall \beta \\ a^T X\beta &= c^T \beta \qquad \forall \beta \end{aligned}$$

which means that $a^T X = c^T$. This implies that c must be in the range space of X^T which in turn implies that c is also in the range space of $X^T X$ which means there exists a λ such that $c = X^T X \lambda$ so:

$$c^T \hat{\beta} = \lambda^T X^T X \hat{\beta} = \lambda^T X^T y$$

Now we can show that the least squares estimator has the minimum variance — pick an arbitrary estimate $a^T y$ and compute its variance:

$$
\begin{aligned}
\text{var}\,(a^T y) &= \text{var}\,(a^T y - c^T \hat{\beta} + c^T \hat{\beta}) \\
&= \text{var}\,(a^T y - \lambda^T X^T y + c^T \hat{\beta}) \\
&= \text{var}\,(a^T y - \lambda^T X^T y) + \text{var}\,(c^T \hat{\beta}) + 2\text{cov}(a^T y - \lambda^T X^T y, \lambda^T X^T y)
\end{aligned}
$$

but

$$
\begin{aligned}
\text{cov}(a^T y - \lambda^T X^T y, \lambda^T X^T y) &= (a^T - \lambda^T X^T)\sigma^2 I X \lambda \\
&= (a^T X - \lambda^T X^T X)\sigma^2 I \lambda \\
&= (c^T - c^T)\sigma^2 I \lambda = 0
\end{aligned}
$$

so

$$\text{var}\,(a^T y) = \text{var}\,(a^T y - \lambda^T X^T y) + \text{var}\,(c^T \hat{\beta})$$

Now since variances cannot be negative, we see that:

$$\text{var}\,(a^T y) \geq \text{var}\,(c^T \hat{\beta})$$

In other words, $c^T \hat{\beta}$ has minimum variance. It now remains to show that it is unique. There will be equality in the above relationship if $\text{var}\,(a^T y - \lambda^T X^T y) = 0$ which would require that $a^T - \lambda^T X^T = 0$ which means that $a^T y = \lambda^T X^T y = c^T \hat{\beta}$. So equality occurs only if $a^T y = c^T \hat{\beta}$ so the estimator is unique. This completes the proof.

The Gauss–Markov theorem shows that the least squares estimate $\hat{\beta}$ is a good choice, but it does require that the errors are uncorrelated and have equal variance. Even if the errors behave, but are nonnormal, then nonlinear or biased estimates may work better. So this theorem does not tell one to use least squares all the time; it just strongly suggests it unless there is some strong reason to do otherwise. Situations where estimators other than ordinary least squares should be considered are:

1. When the errors are correlated or have unequal variance, generalized least squares should be used. See Section 8.1.

2. When the error distribution is long-tailed, then robust estimates might be used. Robust estimates are typically not linear in y. See Section 8.4.

3. When the predictors are highly correlated (collinear), then biased estimators such as ridge regression might be preferable. See Chapter 11.

2.9   Goodness of Fit

It is useful to have some measure of how well the model fits the data. One common choice is R^2, the so-called *coefficient of determination* or *percentage of variance explained*:

$$R^2 = 1 - \frac{\Sigma(\hat{y}_i - y_i)^2}{\Sigma(y_i - \bar{y})^2} = 1 - \frac{\text{RSS}}{\text{Total SS(Corrected for Mean)}}$$

Its range is $0 \leq R^2 \leq 1$ — values closer to 1 indicating better fits. For simple linear regression $R^2 = r^2$ where r is the correlation between x and y. An equivalent definition is:

$$R^2 = \frac{\Sigma(\hat{y}_i - \bar{y})^2}{\Sigma(y_i - \bar{y})^2}$$

or

$$R^2 = \text{cor}^2(\hat{y}, y)$$

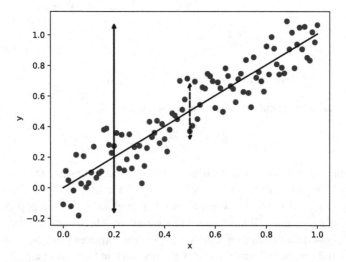

Figure 2.2   When x is not known, the best predictor of y is $\bar{y}$ and the variation is denoted by the solid arrow. When x is known, we can predict y more accurately by the dashed arrow. R^2 is related to the ratio of these two variances.

The graphical intuition behind R^2 is seen in Figure 2.2. Suppose you want to predict y. If you do not know x, then your best prediction is $\bar{y}$, but the variability in this prediction is high. If you do know x, then your prediction will be given by the regression fit. This prediction will be less variable provided there is some relationship between x and y. R^2 is one minus the ratio of the sum of squares for these two predictions. Thus for perfect predictions the ratio will be zero and R^2 will be one.

The code for producing Figure 2.2, which we include for completeness is:

```
x = np.linspace(0,1,101)
np.random.seed(123)
y = x + np.random.normal(0,0.1,101)
plt.scatter(x,y,alpha=0.75)
plt.xlabel("x")
plt.ylabel("y")
beta1, beta0 = np.polyfit(x,y,1)
plt.plot([0,1],[beta0,beta0+beta1],"k-")
plt.annotate("",xy=(0.2, min(y)),xytext=(0.2, max(y)),
    arrowprops=dict(arrowstyle="<->",lw=2))
plt.annotate("",xy=(0.5, 0.3),   xytext=(0.5, 0.7),
    arrowprops=dict(arrowstyle="<->",linestyle="--",lw=2))
```

Note the use of `polyfit` from the numpy package which can be used to fit polynomials for univariate x using least squares. In this case, the degree is one and we only need the fitted coefficients to plot the line.

What is a good value of R^2? It depends on the area of application. In the biological and social sciences, variables tend to be more weakly correlated and there is a lot of noise. We would expect lower values for R^2 in these areas — a value of, say, 0.6 might be considered good. In physics and engineering, where most data come from closely controlled experiments, we typically expect to get much higher R^2s and a value of 0.6 would be considered low. Some experience with the particular area is necessary for you to judge your R^2s well.

It is a mistake to rely on R^2 as a sole measure of fit. Consider the four datasets depicted in Figure 2.3 which may be constructed with:

```
df = sns.load_dataset("anscombe")
sns.lmplot(x="x", y="y", col="dataset",
    data=df, col_wrap=2, ci=None)
```

These datasets were constructed by (Anscombe 1973) and can be found as an example data in the seaborn package which we also use to make the plot. We may compute the R^2 for each of these four datasets as the squared correlation using:

```
df.groupby("dataset").corr().iloc[0::2,-1]**2
```

dataset		
I	x	0.666542
II	x	0.666242
III	x	0.666324
IV	x	0.666707

We see these are essentially the same and yet the points are configured very differently in the four datasets. The first plot on the upper left shows an ordinary relationship while on the right the relationship is nonlinear. On the lower left, the fit looks good except for one outlier demonstrating how sensitive R^2 is to a few extreme values. On the lower right, we see no relationship between the two variables but one outlier leads to a sizable R^2. Another lesson we may take from this example is that correlation can be a poor measure of the relationship between two variables.

An alternative measure of fit is $\hat{\sigma}$. This quantity is directly related to the standard errors of estimates of β and predictions. The advantage is that $\hat{\sigma}$ is measured in the units of the response and so may be directly interpreted in the context of the particular dataset. This may also be a disadvantage in that one must understand the practical significance of this measure whereas R^2, being unitless, is easy to understand. The

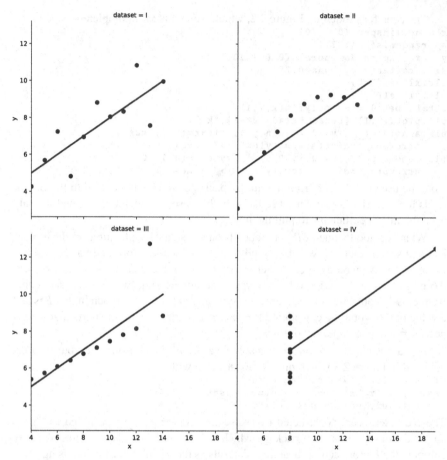

Figure 2.3   Four simulated datasets where R^2 is about 0.65. The plot on the upper left is well behaved for R^2. In the plot on the upper right the relationship is not linear. In the lower two plots, a single outlier has a large effect, lessening the relationship on the left and creating it on the right.

regression summary returns both values and it is worth paying attention to both of them.

2.10   Identifiability

The least squares estimate is the solution to the normal equations:

$$X^T X \hat{\beta} = X^T y$$

where X is an $n \times p$ matrix. If $X^T X$ is singular and cannot be inverted, then there will be infinitely many solutions to the normal equations and $\hat{\beta}$ is at least partially unidentifiable. Unidentifiability will occur when X is not of full rank — that is when its columns are linearly dependent. With observational data, unidentifiability is usually caused by some oversight. Here are some examples:

1. A person's weight is measured both in pounds and kilos and both variables are entered into the model. One variable is just a multiple of the other.

2. For each individual we record the number of years of preuniversity education, the number of years of university education and also the total number of years of education and put all three variables into the model. There is an exact linear relation among the variables.

3. We have more variables than cases, that is, $p > n$. When $p = n$, we may perhaps estimate all the parameters, but with no degrees of freedom left to estimate any standard errors or do any testing. Such a model is called *saturated*. When $p > n$, then the model is sometimes called *supersaturated*. Such models are considered in large-scale screening experiments used in product design and manufacture and in bioinformatics where there are more genes than individuals tested, but there is no hope of uniquely estimating all the parameters in such a model. Different approaches are necessary.

Such problems can be avoided by paying attention. Identifiability is more of an issue in designed experiments. Consider a simple two-sample experiment, where the treatment observations are $y_1, \ldots, y_n$ and the controls are $y_{n+1}, \ldots, y_{m+n}$. Suppose we try to model the response by an overall mean μ and group effects α_1 and α_2:

$$y_j = \mu + \alpha_i + \varepsilon_j \qquad i = 1, 2 \quad j = 1, \ldots, m+n$$

$$
\begin{pmatrix} y_1 \\ \ldots \\ y_n \\ y_{n+1} \\ \ldots \\ y_{m+n} \end{pmatrix}
=
\begin{pmatrix} 1 & 1 & 0 \\ & \ldots & \\ 1 & 1 & 0 \\ 1 & 0 & 1 \\ . & . & . \\ 1 & 0 & 1 \end{pmatrix}
\begin{pmatrix} \mu \\ \alpha_1 \\ \alpha_2 \end{pmatrix}
+
\begin{pmatrix} \varepsilon_1 \\ \ldots \\ \ldots \\ \ldots \\ \varepsilon_{m+n} \end{pmatrix}
$$

Now although X has three columns, it has only rank two — $(\mu, \alpha_1, \alpha_2)$ are not identifiable and the normal equations have infinitely many solutions. We can solve this problem by imposing some constraints, $\mu = 0$ or $\alpha_1 + \alpha_2 = 0$, for example.

Statistics packages handle nonidentifiability differently. In the regression case above, some may return error messages and some may fit models because rounding error may remove the exact identifiability. In other cases, constraints may be applied but these may be different from what you expect. Here is an example. Suppose we create a new variable for the Galápagos dataset — the difference in area between the island and its nearest neighbor:

```
galapagos['Adiff'] = galapagos['Area'] - galapagos['Adjacent']
```

and add that to the model. We divide the model specification from the fitting (because we are going to use a second fitting method soon).

```
lmodform = smf.ols(
    'Species ~ Area+Elevation+Nearest+Scruz+Adjacent+Adiff',
    galapagos)
```

```
lmod = lmodform.fit()
lmod.sumary()
          coefs stderr tvalues pvalues
Intercept  7.068 19.154    0.37  0.7154
Area      -0.041  0.018   -2.24  0.0349
Elevation  0.319  0.054    5.95  0.0000
Nearest    0.009  1.054    0.01  0.9932
Scruz     -0.241  0.215   -1.12  0.2752
Adjacent  -0.058  0.016   -3.51  0.0018
Adiff      0.017  0.007    2.34  0.0279
```

```
n=30 p=6 Residual SD=60.975 R-squared=0.77
Warning: Strong collinearity - design may be singular
```

The Moore-Penrose method is able to compute parameter estimates even in the presence of a singularity but this is problematic. If we examine the smallest eigenvalue,

```
lmod.eigenvals[-1]
2.4489090710071553e-24
```

we find a very small value indicating singularity. This is because the rank of the design matrix X is six, which is less than its seven columns. In most cases, the cause of identifiability can be revealed with some thought about the variables, but, failing that, an eigendecomposition of $X^T X$ will reveal the linear combination(s) that gave rise to the unidentifiability — see Section 11.1.

Alternatively, we can use the QR decomposition, as described in Section 2.7:

```
lmod = lmodform.fit(method="qr")
lmod.sumary()
```

```
                        coefs stderr tvalues pvalues
Intercept               7.887 19.197    0.41  0.6848
Area      -18,382,410,658,694.098    nan     nan     nan
Elevation               0.318  0.054    5.91  0.0000
Nearest                 0.171  1.056    0.16  0.8726
Scruz                  -0.269  0.216   -1.25  0.2240
Adjacent   18,382,410,658,694.004    nan     nan     nan
Adiff      18,382,410,658,694.074    nan     nan     nan
```

```
n=30 p=6 Residual SD=61.110 R-squared=0.76
Warning: Strong collinearity - design may be singular
```

You can clearly see that something has gone badly wrong (and I have omitted the many warning messages that were also generated). There is a silver lining to this particular cloud since not only are we alerted to the problem, we can see which three variable are involved in the linear dependency.

For interactive use, where you plan to interpret the model and perform inference using it, the use of the Moore Penrose method has an element of danger. If one misses the warning signs, you will be off on the wrong track. On the other hand, the model will still produce valid and relatively stable predictions. For models that are fit without human oversight, this may be what is required. The advantage of the QR method is that failure will be clearly apparent and the necessary amputation of the offending variables will prevent the infection from spreading.

Lack of identifiability is obviously a problem, but once spotted, we can take evasive action. More problematic are cases where we are close to unidentifiability. To demonstrate this, suppose we add a small random perturbation to the third decimal place of Adiff by adding a random variate from $U[-0.005, 0.005]$ where U denotes the uniform distribution. Random numbers are by nature random so results are not exactly reproducible. However, you can make the numbers come out the same every time by setting the seed on the random number generator using np.random.seed(). I have done this here so you will not wonder why your answers are not exactly the same as mine, but it is not strictly necessary.

```
np.random.seed(123)
galapagos['Adiffe'] = galapagos['Adiff'] + \
    (np.random.rand(30) -0.5)*0.001
```

and now refit the model:

```
lmod = smf.ols(
    'Species ~ Area+Elevation+Nearest+Scruz+Adjacent+Adiffe',
    galapagos).fit()
lmod.sumary()
```

	coefs	stderr	tvalues	pvalues
Intercept	10.492	19.726	0.53	0.5999
Area	-45,266.870	54,925.786	-0.82	0.4183
Elevation	0.313	0.055	5.73	0.0000
Nearest	-0.065	1.065	-0.06	0.9518
Scruz	-0.234	0.217	-1.08	0.2914
Adjacent	45,266.770	54,925.783	0.82	0.4183
Adiffe	45,266.846	54,925.785	0.82	0.4183

n=30 p=7 Residual SD=61.387 R-squared=0.77

Notice that now all parameters are estimated, but some estimates and standard errors are very large because we cannot estimate them in a stable way. We set up this problem so we know the cause, but in general we need to be able to identify such situations. We do this in Section 7.3.

2.11 Orthogonality

Orthogonality is a useful property because it allows us to more easily interpret the effect of one predictor without regard to another. Suppose we can partition X in two, $X = [X_1|X_2]$ such that $X_1^T X_2 = 0$. So now:

$$Y = X\beta + \varepsilon = X_1\beta_1 + X_2\beta_2 + \varepsilon$$

and

$$X^T X = \begin{pmatrix} X_1^T X_1 & X_1^T X_2 \\ X_2^T X_1 & X_2^T X_2 \end{pmatrix} = \begin{pmatrix} X_1^T X_1 & 0 \\ 0 & X_2^T X_2 \end{pmatrix}$$

which means:

$$\hat{\beta}_1 = (X_1^T X_1)^{-1} X_1^T y \qquad \hat{\beta}_2 = (X_2^T X_2)^{-1} X_2^T y$$

Notice that $\hat{\beta}_1$ will be the same regardless of whether X_2 is in the model or not (and vice versa). So we can interpret the effect of X_1 without a concern for X_2. Unfortunately, the decoupling is not perfect. Suppose we wish to test whether $\beta_1 = 0$. We have RSS/$df = \hat{\sigma}^2$ that will be different depending on whether X_2 is included in the model or not, but the difference in the test statistic is not liable to be as large as in nonorthogonal cases.

If the covariance between vectors x_1 and x_2 is zero, then $\sum_j (x_{j1} - \bar{x}_1)(x_{j2} - \bar{x}_2) = 0$. This means that if we center the predictors, a covariance of zero implies orthogonality. As can be seen in the second example in Section 2.5, we can center the predictors without essentially changing the model provided we have an intercept term.

Orthogonality is a desirable property, but will only occur when X is chosen by the experimenter. It is a feature of a good design. In observational data, we do not have direct control over X and this is the source of many of the interpretational difficulties associated with nonexperimental data.

Here is an example of an experiment to determine the effects of column temperature, gas/liquid ratio and packing height in reducing the unpleasant odor of a chemical product that was sold for household use. Read the data in and display with:

```
import faraway.datasets.odor
odor = faraway.datasets.odor.load()
odor.head()
   odor  temp  gas  pack
0    66    -1   -1     0
1    39     1   -1     0
2    43    -1    1     0
3    49     1    1     0
4    58    -1    0    -1
```

The three predictors have been transformed from their original scale of measurement, for example, temp = (Fahrenheit-80)/40 so the original values of the predictor were 40, 80 and 120. The data is presented in John (1971) and give an example of a *central composite* design. We compute the covariance of the predictors:

```
odor.iloc[:,1:].cov()
          temp        gas      pack
temp  0.571429  0.000000  0.000000
gas   0.000000  0.571429  0.000000
pack  0.000000  0.000000  0.571429
```

The matrix is diagonal. Even if temp was measured in the original Fahrenheit scale, the matrix would still be diagonal, but the entry in the matrix corresponding to temp would change. Now fit a model and output the parameter estimates:

```
lmod = smf.ols('odor ~ temp + gas + pack', odor).fit()
lmod.params
Intercept    15.200
temp        -12.125
gas         -17.000
pack        -21.375
```

We can get the covariances as:

```
odor.iloc[:,1:].cov()
           Intercept   temp    gas   pack
Intercept      86.46   -0.0   -0.0   -0.0
temp           -0.00  162.1   -0.0    0.0
gas            -0.00   -0.0  162.1    0.0
pack           -0.00    0.0    0.0  162.1
```

We see that, as expected, the pairwise covariance of all the coefficients is zero. Notice that the variances for the coefficients are equal due to the balanced design. Now drop one of the variables:

```
lmod = smf.ols('odor ~ gas + pack', odor).fit()
```

```
lmod.params
Intercept    15.200
gas         -17.000
pack        -21.375
```
The coefficients themselves do not change, but the residual SE does change slightly, which causes small changes in the SEs of the coefficients, t-statistics and p-values, but nowhere near enough to change our qualitative conclusions.

Exercises

1. The dataset teengamb concerns a study of teenage gambling in Britain. Treat gambling as the response and the sex, status, income and verbal score as predictors.

   (a) Fit the linear model and present the output.

   (b) What percentage of variation in the response is explained by these predictors?

   (c) Which observation has the largest (positive) residual? Give the case number. Use the idxmax() function from pandas.

   (d) Compute the mean and median of the residuals. Use the mean() and median() functions from pandas.

   (e) Compute the correlation of the residuals with the fitted values. Will it always be this value? Use the corr() function from pandas.

   (f) Compute the correlation of the residuals with the income. Will it always be this value?

   (g) For all other predictors held constant, what would be the difference in predicted expenditure on gambling for a male compared to a female?

2. The dataset uswages is drawn as a sample from the Current Population Survey in 1988.

   (a) Fit a model with weekly wages as the response and years of education and experience as predictors. Show the output.

   (b) What are the minimum values of wages, education and experience seen in the data? Use the min() function from pandas.

   (c) What is the predicted wage for a worker with no education and no experience? Comment.

   (d) Verify by calculation that the squared correlation between the fitted values and the response is equal to R^2.

   (e) In view of the difficulty of modeling uneducated, inexperienced men, we might try omitting the intercept term from the model. We can accomplish this with a model formula like y ~ -1 + x. Fit the same predictors but without the intercept term. What is the reported value of R^2? How does this compare to the squared correlation between the fitted values and the response?

   (f) Compare the residual sum of squares for the two models we have fitted so far. Which model fits best?

(g) If education were increased by one, holding experience constant, what would be the change in the predicted wage?

(h) Fit a model with log(wages) as the response and education and experience as predictors. Compare the residual standard error and R^2 for this model with those seen in the first model. Which model do you think is best?

(i) If education were increased by one, holding experience constant, what would be the change in the predicted wage using this logged model?

(j) Add the predictors ne, mw, we and so into the model. Fit the model using the default Moore-Penrose method and the QR method. Why are the results different? Compute the smallest eigenvalue of the $X^T X$ matrix and comment.

(k) Compute the row sum (i.e., over individual men) for ne, mw, we and so. What value does this take for all men? What relevance does this have to the previous question?

3. The dataset prostate comes from a study on 97 men with prostate cancer who were due to receive a radical prostatectomy.

(a) Fit a model with lpsa as the response and lcavol as the predictor. Record the residual standard error and the R^2.

(b) Now add lweight, svi, lbph, age, lcp, pgg45 and gleason to the model sequentially one at a time. For each model, record the residual standard error and the R^2.

(c) Plot the trends in these two statistics. Comment on the results - do they change monotonically as the number of predictors increases?

4. Using the prostate data, plot lpsa on the vertical and lcavol on the horizontal. Fit the regressions of lpsa on lcavol and lcavol on lpsa. Display both regression lines on the plot. At what point do the two lines intersect?

5. Thirty samples of cheddar cheese were analyzed for their content of acetic acid, hydrogen sulfide and lactic acid. Each sample was tasted and scored by a panel of judges and the average taste score produced. Use the cheddar data to answer the following:

(a) Fit a regression model with taste as the response and the three chemical contents as predictors. Report the values of the regression coefficients.

(b) Compute the correlation between the fitted values and the response. Square it. Identify where this value appears in the regression output.

(c) Fit the same regression model but without an intercept term. What is the value of R^2 reported in the output? Compute a more reasonable measure of the goodness of fit for this example.

(d) Compute the regression coefficients from the original fit using the QR decomposition showing your code.

6. An experiment was conducted to determine the effect of four factors on the resistivity of a semiconductor wafer. The data is found in wafer where each of the four factors is coded as − or + depending on whether the low or the high setting for that factor was used. Fit the linear model resist ~ x1 + x2 + x3 + x4.

(a) Extract the X matrix using the following code:

```
import patsy
X = patsy.dmatrix("x1 + x2 + x3 + x4",
    data=wafer, return_type = 'dataframe')
X
```

Examine this to determine how the low and high levels have been coded in the model.

(b) Compute the correlation in the X matrix. Why are there some missing values in the matrix?

(c) What difference in resistance is expected when moving from the low to the high level of x1?

(d) Refit the model without x4 and examine the regression coefficients and standard errors? What stayed the same as the original fit and what changed?

(e) Explain how the change in the regression coefficients is related to the correlation matrix of X.

7. An experiment was conducted to examine factors that might affect the height of leaf springs in the suspension of trucks. The data may be found in truck.

(a) The five factors in the experiment are set to $-$ and $+$ but it will be more convenient for us to use -1 and $+1$. This can be achieved for the first factor by:

```
truck.B = np.where(truck.B == '-',-1,+1)
```

Repeat for the other four factors.

(b) Fit a linear model for the height in terms of the five factors. Report on the value of the regression coefficients.

(c) Fit a linear model using just factors B, C, D and E and report the coefficients. How do these compare to the previous question? Show how we could have anticipated this result by examining the X matrix.

(d) Construct a new predictor called A which is set to B+C+D+E. Fit a linear model with the predictors A, B, C, D, E and O. What is the difficulty with this model?

(e) Fit the previous model using the QR decomposition method and comment on the results.

Chapter 3

Inference

In this chapter we show how to make hypothesis tests and construct confidence intervals. These inferential methods are the building blocks for drawing conclusions using models and data.

If you simply wish to estimate the parameters β, it is not essential to assume any distributional form for the errors ε. However, if we want to make any confidence intervals or perform any hypothesis tests using the most commonly used methods, we will need to do this. Now we are going to assume that the errors are normally distributed. Often this is quite reasonable but we will discuss later how we can check this assumption and how we might deal with data where the normality assumption is not justifiable.

If we have chosen to use least squares estimation, we should already have assumed that the errors are independent and identically distributed (i.i.d.) with mean 0 and variance σ^2. We can represent this compactly as $\varepsilon \sim N(0, \sigma^2 I)$ where the N here represents a multivariate normal, the mean is a vector of zeroes and the covariance matrix is $\sigma^2 I$. Now since $y = X\beta + \varepsilon$, we have $y \sim N(X\beta, \sigma^2 I)$ which is a compact description of the regression model. From this we find, using the fact that linear combinations of normally distributed values are also normal, that:

$$\hat{\beta} = (X^T X)^{-1} X^T y \sim N(\beta, (X^T X)^{-1} \sigma^2)$$

3.1 Hypothesis Tests to Compare Models

Given several predictors for a response, we might wonder whether all are needed. Consider a larger model, Ω, and a smaller model, ω, which consists of a subset of the predictors that are in Ω. If there is not much difference in the fit, we would prefer the smaller model on the principle that simpler explanations are preferred. On the other hand, if the fit of the larger model is appreciably better, we will prefer it. We will take ω to represent the null hypothesis and Ω to represent the alternative. A geometrical view of the problem may be seen in Figure 3.1.

If $RSS_\omega - RSS_\Omega$ is small, then the fit of the smaller model is almost as good as the larger model and so we would prefer the smaller model on the grounds of simplicity. On the other hand, if the difference is large, then the superior fit of the larger model would be preferred. This suggests that something like:

$$\frac{RSS_\omega - RSS_\Omega}{RSS_\Omega}$$

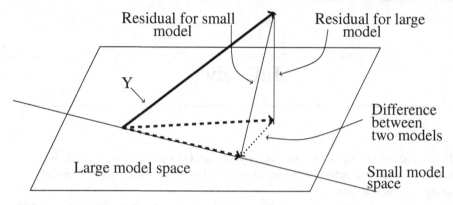

Figure 3.1 Geometric view of the comparison between big model, Ω, and small model, ω. The squared length of the residual vector for the big model is RSS_Ω, while that for the small model is RSS_ω. By the Pythagoras theorem, the squared length of the vector connecting the two fits is $RSS_\omega - RSS_\Omega$. A small value for this indicates that the small model fits almost as well as the large model and thus might be preferred due to its simplicity.

would be a potentially good test statistic where the denominator is used for scaling purposes.

As it happens, the same test statistic arises from the likelihood-ratio testing approach. We give an outline of the development: If $L(\beta, \sigma|y)$ is the likelihood function, then the likelihood-ratio statistic is:

$$\frac{\max_{\beta, \sigma \in \Omega} L(\beta, \sigma|y)}{\max_{\beta, \sigma \in \omega} L(\beta, \sigma|y)}$$

The test should reject if this ratio is too large. Working through the details, we find that for each model:

$$L(\hat{\beta}, \hat{\sigma}|y) \propto \hat{\sigma}^{-n}$$

which after some manipulation gives us a test that rejects if:

$$\frac{RSS_\omega - RSS_\Omega}{RSS_\Omega} > \text{a constant}$$

which is the same statistic suggested by the geometric view. Now suppose that the dimension (or number of parameters) of Ω is p and the dimension of ω is q, then we can use some more scaling to get an F-statistic which has an F-distribution under the null hypothesis:

$$F = \frac{(RSS_\omega - RSS_\Omega)/(p - q)}{RSS_\Omega/(n - p)} \sim F_{p-q, n-p}$$

Details of the derivation of this statistic may be found in more theoretically oriented texts such as Sen and Srivastava (1990).

Thus we would reject the null hypothesis if $F > F_{p-q,n-p}^{(\alpha)}$. The degrees of freedom of a model are (usually) the number of observations minus the number of parameters so this test statistic can also be written:

$$F = \frac{(\mathrm{RSS}_\omega - \mathrm{RSS}_\Omega)/(df_\omega - df_\Omega)}{\mathrm{RSS}_\Omega/df_\Omega}$$

where $df_\Omega = n - p$ and $df_\omega = n - q$. The same test statistic applies not just to when ω is a subset of Ω, but also to a subspace. This test is very widely used in regression and analysis of variance. When it is applied in different situations, the form of test statistic may be reexpressed in various ways. The beauty of this approach is you only need to know the general form. In any particular case, you just need to figure out which models represent the null and alternative hypotheses, fit them and compute the test statistic. The test is very versatile.

3.2 Testing Examples

Test of all the predictors

Are any of the predictors useful in predicting the response? Let the full model (Ω) be $y = X\beta + \varepsilon$ where X is a full-rank $n \times p$ matrix and the reduced model (ω) be $y = \mu + \varepsilon$. We estimate μ by $\bar{y}$. We write the null hypothesis as:

$$H_0 : \beta_1 = \ldots \beta_{p-1} = 0$$

Now $\mathrm{RSS}_\Omega = (y - X\hat{\beta})^T (y - X\hat{\beta}) = \hat{\varepsilon}^T \hat{\varepsilon}$, the residual sum of squares for the full model, while $\mathrm{RSS}_\omega = (y - \bar{y})^T (y - \bar{y}) = TSS$, which is called the total sum of squares corrected for the mean. The F-statistic is:

$$F = \frac{(TSS - RSS)/(p-1)}{RSS/(n-p)}$$

We would now refer to $F_{p-1,n-p}$ for a critical value or a p-value. Large values of F would indicate rejection of the null. Traditionally, the information in the above test is presented in an *analysis of variance* or ANOVA table. Most computer packages produce a variant on this. See Table 3.1 for a typical layout. It is not really necessary to specifically compute all the elements of the table. As the originator of the table, Fisher said in 1931, it is "nothing but a convenient way of arranging the arithmetic." Since he had to do his calculations by hand, the table served a necessary purpose, but it is not essential now.

Source	Deg. of Freedom	Sum of Squares	Mean Square	F
Regression	$p-1$	SS_{reg}	$SS_{reg}/(p-1)$	F
Residual	$n-p$	RSS	$RSS/(n-p)$	
Total	$n-1$	TSS		

Table 3.1   Analysis of variance table.

A failure to reject the null hypothesis is not the end of the game — you must still investigate the possibility of nonlinear transformations of the variables and of outliers which may obscure the relationship. Even then, you may just have insufficient data to demonstrate a real effect, which is why we must be careful to say "fail to reject" the null rather than "accept" the null. It would be a mistake to conclude that no real relationship exists. This issue arises when a pharmaceutical company wishes to show that a proposed generic replacement for a brand-named drug is equivalent. It would not be enough in this instance just to fail to reject the null. A higher standard would be required.

When the null is rejected, this does not imply that the alternative model is the best model. We do not know whether all the predictors are required to predict the response or just some of them. Other predictors might also be added or existing predictors transformed or recombined. Either way, the overall F-test is just the beginning of an analysis and not the end.

Let's illustrate this test using the Galápagos Islands dataset introduced in the last chapter. As before, we load all the packages we will use in this chapter:

```
import pandas as pd
import numpy as np
import matplotlib.pyplot as plt
import scipy as sp
import statsmodels.api as sm
import statsmodels.formula.api as smf
import faraway.utils
```

We fit the same model as before with the number of species as the response and the geographic variables as predictors.

```
import faraway.datasets.galapagos
galapagos = faraway.datasets.galapagos.load()
lmod = smf.ols('Species ~ Area + Elevation + Nearest + \
        Scruz + Adjacent', galapagos).fit()
```

We can obtain the F-statistic and corresponding p-value from the model object.

```
lmod.fvalue, lmod.f_pvalue
  (15.69, 6.84e-07)
```

From this we see directly the result of the test of whether any of the predictors have significance in the model — that is, whether $\beta_1 = \beta_2 = \beta_3 = \beta_4 = \beta_5 = 0$. Since the p-value of 6.84×10^{-7} is so small, this null hypothesis is rejected.

Another way to do it:

```
lmodr = smf.ols('Species ~ 1', galapagos).fit()
lmod.compare_f_test(lmodr)
  (15.69, 6.84e-07, 5.0)
```

Just for verification purposes, we can also do it directly using the F-testing formula. The total sum of squares (which comes from fitting the null model) and the residual sum of squares are:

```
lmod.centered_tss, lmod.ssr
  (381081, 89231)
```

the respective degrees of freedom are:

```
lmod.df_model, lmod.df_resid
  (5, 24)
```

and the mean squares (dividing the sum of squares by the degrees of freedom are)

```
lmod.mse_model, lmod.mse_resid
  (58370, 3718)
```

Hence the F-statistic is:
```
lmod.mse_model/ lmod.mse_resid
15.69
```
and the *p*-value is
```
1-sp.stats.f.cdf(lmod.fvalue, lmod.df_model, lmod.df_resid)
6.84e-07
```
This is just for demonstration purposes — in practice, we will use the first method of calculation.

Testing one predictor

Can one particular predictor be dropped from the model? The null hypothesis is then $H_0 : \beta_i = 0$. Let Ω be the model with all the predictors of interest which has p parameters and let ω be the model with all the same predictors except predictor i. Let's test whether Area can be dropped from the full model by testing the hypothesis that the corresponding parameter is zero. Using the general method, we fit a model without Area and obtain this:
```
lmods = smf.ols('Species ~ Elevation + Nearest + \
        Scruz + Adjacent', galapagos).fit()
sm.stats.anova_lm(lmods,lmod)
   df_resid        ssr  df_diff     ss_diff         F    Pr(>F)
0      25.0  93469.08399     0.0         NaN       NaN       NaN
1      24.0  89231.36633     1.0  4237.71766  1.139792  0.296318
```
The NaNs in the output can be ignored. The *p*-value of 0.296 indicates that the null hypothesis cannot be rejected here.

An alternative approach is to use a *t*-statistic for testing the hypothesis:

$$t_i = \hat{\beta}_i / se(\hat{\beta}_i)$$

and check for significance using a *t*-distribution with $n - p$ degrees of freedom. It can be shown that t_i^2 is equal to the appropriate F-statistic computed using the method shown above. We can see the *t*-statistic and *p*-value in usual regression summary output:
```
lmod.sumary()
           coefs  stderr  tvalues  pvalues
Intercept  7.068  19.154    0.37   0.7154
Area      -0.024   0.022   -1.07   0.2963
Elevation  0.319   0.054    5.95   0.0000
Nearest    0.009   1.054    0.01   0.9932
Scruz     -0.241   0.215   -1.12   0.2752
Adjacent  -0.075   0.018   -4.23   0.0003

n=30 p=6 Residual SD=60.975 R-squared=0.77
```
We can verify that this is indeed the same result. Of course, it is easier to obtain the summary output, so we will use this method from now on.

It is important to be precise about what hypothesis is being tested. For example, consider another test of $H_0 : \beta_{Area} = 0$:
```
lmods = smf.ols('Species ~ Elevation + Nearest + Scruz',
        galapagos).fit()
sm.stats.anova_lm(lmods,lmod)
   df_resid          ssr  df_diff       ss_diff         F   Pr(>F)
0      26.0  158291.628568     0.0           NaN       NaN      NaN
1      24.0   89231.366330     2.0  69060.262238  9.287352  0.00103
```

We see a very different p-value of 0.001 which changes the conclusion of the hypothesis test. The null is firmly rejected.

From this we can see that it is very important that we specify which other predictors are included in the models specifying the null and alternative. It is not sufficient simply to state the null hypothesis as $H_0 : \beta_{\text{Area}} = 0$.

Now we may ask what is the effect of area on the number of species. We postpone discussion until Chapter 5, but we can at least see that the question will have no simple answer.

Testing a pair of predictors

Suppose we were interested in whether the area of either the current island or the adjacent island had any relation to the response. This corresponds to a null hypothesis of $H_0 : \beta_{\text{Area}} = \beta_{\text{Adjacent}} = 0$ where we shall also specify that other three predictors are included in the model. We can test this by fitting a model without these two terms and constructing the F-test:

```
lmods = smf.ols('Species ~ Elevation + Nearest + Scruz',
        galapagos).fit()
sm.stats.anova_lm(lmods,lmod)
   df_resid          ssr  df_diff       ss_diff         F   Pr(>F)
0      26.0  158291.628568      0.0          NaN       NaN      NaN
1      24.0   89231.366330      2.0  69060.262238  9.287352  0.00103
```

The null hypothesis is rejected because the p-value is small. This tells us that such a simplification to remove these two predictors is not justifiable.

Now you may wonder whether we could have divined the same result from looking at the regression summary output where the corresponding p-values for the two terms are 0.3 and 0.0003. Immediately, we have the problem of making a decision based on two p-values rather than one. There is no simple way to combine these. Furthermore, each of these p-values corresponds to a test where the other predictor is included in the model. So in short, if you want to test two (or more) predictors, you need to use the single F-test. You cannot reliably use the two (in this case) t-tests.

Testing a subspace

Some tests cannot be expressed simply in terms of the inclusion or exclusion of subsets of predictors. Consider an example where we test whether the areas of the current and adjacent island can be added together and used in place of the two separate predictors. Such a test can be expressed as the hypothesis that:

$$H_0 : \beta_{\text{Area}} = \beta_{\text{Adjacent}}$$

The model corresponding to this null hypothesis represents a linear subspace of the full model. We can test this by specifying the null model and applying the F-test procedure:

```
lmods = smf.ols('Species ~ I(Area+Adjacent) + \
        Elevation + Nearest + Scruz', galapagos).fit()
sm.stats.anova_lm(lmods,lmod)
   df_resid          ssr  df_diff       ss_diff         F   Pr(>F)
0      25.0  109591.120801      0.0          NaN       NaN      NaN
1      24.0   89231.366330      1.0  20359.754471  5.476035  0.027926
```

The function I() ensures that the argument is evaluated (in this case actual addition) rather than interpreted as part of the model formula. The p-value of 0.028 indicates that the null can be rejected here and the proposed simplification to a single combined area predictor is not justifiable.

Another example of subspace testing occurs when we want to test whether a parameter can be set to a particular value. For example:

$$H_0 : \beta_{\text{Elevation}} = 0.5$$

This specifies a particular subspace of the full model. We can set a fixed term in the regression equation using an *offset*. We fit this model and compare it to the full:

```
lmod = smf.glm('Species ~ Area + Elevation + Nearest + \
        Scruz + Adjacent', galapagos).fit()
lmods = smf.glm('Species ~ Area + Nearest + Scruz + \
        Adjacent', offset=(0.5*galapagos['Elevation']),
        data=galapagos).fit()
fstat = (lmods.deviance-lmod.deviance)/ \
        (lmod.deviance/lmod.df_resid)
pvalue = 1-sp.stats.f.cdf(fstat, 1, lmod.df_resid)
fstat, pvalue
(11.31, 0.0025)
```

We see that the p-value is small and the null hypothesis is rejected. A simpler way to test such point hypotheses is to use a t-statistic:

$$t = (\hat{\beta} - c)/\text{se}(\hat{\beta})$$

where c is the point hypothesis. So in our example the statistic and corresponding p-value are:

```
lmod = smf.ols('Species ~ Area + Elevation + Nearest + \
        Scruz  + Adjacent', galapagos).fit()
tstat=(lmod.params['Elevation']-0.5)/lmod.bse['Elevation']
tstat, 2*sp.stats.t.cdf(tstat, lmod.df_resid)
(-3.36, 0.00257)
```

We can see the p-value is the same as before and if we square the t-statistic

```
tstat**2
11.31
```

we find we get the same F-value as above. This latter approach is preferred in practice since we do not need to fit two models but it is important to understand that it is equivalent to the result obtained using the general F-testing approach.

Tests we cannot do

We cannot test a nonlinear hypothesis like $H_0 : \beta_j\beta_k = 1$ using the F-testing method. We would need to fit a nonlinear model, and that lies beyond the scope of this book.

Also we cannot compare models that are not nested using an F-test. For example, if one model has Area and Elevation as predictors while another has Area, Adjacent and Scruz, we cannot use the F-test. We can ask which model is preferable. Methods for doing this will be discussed in Chapter 10.

A further difficulty can arise when the models we compare use different datasets. This can arise quite easily when values of some variables are missing as different models may use different cases depending on which are complete. F-tests are not directly possible here. Methods for dealing with this are presented in Chapter 13.

3.3   Permutation Tests

The tests we have considered thus far are based on the assumption of normal errors. Arguments based on the central limit theorem mean that even if the errors are not normal, inference based on the assumption of normality can be approximately correct provided the sample size is large enough. Unfortunately, it is not possible to say how large the sample has to be or how close to normality the error distribution has to be before the approximation is satisfactory. Permutation tests offer an alternative that needs no assumption of normality.

We can put a different interpretation on the hypothesis tests we are making. For the Galápagos dataset, we might suppose that if the number of species had no relation to the five geographic variables, then the observed response values would be randomly distributed between the islands without relation to the predictors. The F-statistic is a good measure of the association between the predictors and the response with larger values indicating stronger associations. We might then ask what the chance would be under this assumption that an F-statistic would be observed as large or larger than the one we actually observed. We could compute this exactly by computing the F-statistic for all possible ($n!$) permutations of the response variable and see what proportion exceeds the observed F-statistic. This is a permutation test. If the observed proportion is small, then we must reject the contention that the response is unrelated to the predictors. Curiously, this proportion is estimated by the p-value calculated in the usual way based on the assumption of normal errors thus saving us from the massive task of actually computing the regression on all those computations. See Freedman and Lane (1983) for details.

Let's see how we can apply the permutation test to the Galápagos data. We choose a model with just `Nearest` and `Scruz` so as to get a p-value for the F-statistic that is not too small (and therefore less interesting):

```
lmod = smf.ols('Species ~ Nearest + Scruz',
        galapagos).fit()
```

The function `np.random.permutation()` generates random permutations. We compute the F-statistic for 4000 randomly selected permutations and see what proportion exceeds the F-statistic for the original data:

```
fstats = np.zeros(4000)
np.random.seed(123)
for i in range(0,4000):
    galapagos['ysamp'] = np.random.permutation(
            np.copy(galapagos['Species']))
    lmodi = smf.ols('ysamp ~ Nearest + \
            Scruz', galapagos).fit()
    fstats[i] = lmodi.fvalue
np.mean(fstats > lmod.fvalue)
0.53825
```

This should take only a few seconds on any relatively new computer. We could speed this process up noting that the X-matrix does not change in all 4000 regressions so we could avoid repeating a lot of the calculations. However, when considering such an improvement, we should notice that the slow result will arrive well before we have even typed anything more efficient. Of course, this may not be true for examples with bigger data.

The function np.random.seed ensures that the random numbers used to generate the permutations will come out the same for you if you try this. If you don't do this, you will get a slightly different result each time because of the random selection of the permutations. That's alright — setting the seed is only necessary for exact reproducibility.

Our estimated p-value using the permutation test is 0.54, which is close to the normal theory-based value of 0.55. We could reduce variability in the estimation of the p-value simply by computing more random permutations. Since the permutation test does not depend on the assumption of normality, we might regard it as superior to the normal theory based value. In this case, the results are very similar and not close to any decision boundary. But if there was some crucial difference in the conclusion and there was some evidence of nonnormal errors, then we would prefer the permutation-based test.

Tests involving just one predictor also fall within the permutation test framework. We permute that predictor rather than the response. Let's test the Scruz predictor in the model. We can extract the needed information from:

```
lmod.tvalues[2], lmod.pvalues[2]
```
```
(-1.022 0.3156)
```

Now we perform 4000 permutations of Scruz and check what fraction of the t-statistics exceeds -1.09 in absolute value:

```
tstats = np.zeros(4000)
np.random.seed(123)
for i in range(0, 4000):
    galapagos['ssamp'] = np.random.permutation(galapagos.Scruz)
    lmodi = smf.ols('Species ~ Nearest + ssamp',
                    galapagos).fit()
    tstats[i] = lmodi.tvalues[2]
np.mean(np.fabs(tstats) > np.fabs(lmod.tvalues[2]))
```
```
0.297
```

The outcome is very similar to the observed normal-based p-value of 0.32. Again, in case of serious disagreement, we would prefer the permutation-based result.

The idea of permutation tests works well in conjunction with the principle of random allocation of units in designed experiments. When the values of X really have been randomly assigned to the experimental units which then produce response Y, it is easy to justify a permutation-based testing procedure to check whether there truly is any relation between X and Y. In practice, many will simply use the normal assumption-based tests but this can be done in the knowledge that the permutation tests will tend to agree, provided the assumption is justifiable.

3.4 Sampling

The method of data collection affects the conclusions we can draw. The mathematical model $Y = X\beta + \varepsilon$ describes how the response Y is generated. If we specified β, we could generate Y using a computer simulation, but how can this be related to real data?

For designed experiments, we can view nature as the computer generating the observed responses. We input X and record Y. With no cost or time constraints, we could repeat this as many times as we liked, but in practice we collect a sample of fixed size. Our inference then tells us something about the β underlying this natural process.

For observational studies, we envisage a finite *population* from which we draw the *sample* that is our data. We want to say something about the unknown population value of β, using estimated values $\hat{\beta}$ that are obtained from the sample data. We prefer that the data be a *simple random sample* of the population. We also assume that the size of the sample is a small fraction of the population size. We can also accommodate more complex random sampling designs, but this would require more complex inferential methods.

Sometimes, researchers may try to select a *representative* sample by hand. Quite apart from the obvious difficulties in doing this, the logic behind the statistical inference depends on the sample being random. This is not to say that such studies are worthless, but that it would be unreasonable to apply anything more than descriptive statistical techniques. Confidence in the conclusions from such data is necessarily suspect.

A sample of convenience is where the data are not collected according to a sampling design. In some cases, it may be reasonable to proceed as if the data were collected using a random mechanism. For example, suppose we take the first 400 people from the phone book whose names begin with the letter P. Provided there is no ethnic effect, it may be reasonable to consider this a random sample from the population defined by the entries in the phone book. Here we are assuming the selection mechanism is effectively random with respect to the objectives of the study. The data are as good as random. Other situations are less clear-cut and judgment will be required. Such judgments are easy targets for criticism. Suppose you are studying the behavior of alcoholics and advertise in the media for study subjects. It seems very likely that such a sample will be biased, perhaps in unpredictable ways. In cases such as this, a sample of convenience is clearly biased in which case conclusions must be limited to the sample itself. This situation reduces to the following case.

Sometimes, the sample is the complete population. In this case, one might argue that inference is not required since the population and sample values are one and the same. For the `galapagos` data, the sample is effectively the population or a large and biased proportion thereof. Permutation tests make it possible to give some meaning to the p-value when the sample is the population or for samples of convenience although one has to be clear that the conclusion applies only to the particular sample. Another approach that gives meaning to the p-value when the sample is the population involves the imaginative concept of "alternative worlds" where the sample or population at hand is supposed to have been randomly selected from parallel universes. This argument is more speculative.

3.5   Confidence Intervals for β

Confidence intervals (CIs) provide an alternative way of expressing the uncertainty
in the estimates of β. They take the form:

$$\hat{\beta}_i \pm t_{n-p}^{(\alpha/2)} \text{se}(\hat{\beta}_i)$$

Consider this model for the Galápagos data:

```
lmod = smf.ols('Species ~ Area + Elevation + Nearest + \
        Scruz  + Adjacent', galapagos).fit()
```

We can construct individual 95% CIs for β_{Area} for which we need the 2.5% and
97.5% percentiles of the t-distribution with $30 - 6 = 24$ degrees of freedom. These
are returned using the sp.stats.t.interval where the first argument is the cov-
erage and the second the degrees of freedom:

```
qt = np.array(sp.stats.t.interval(0.95,24))
lmod.params[1] + lmod.bse[1]*qt
array([-0.07179,  0.10105])
```

CIs have a duality with two-sided hypothesis tests. If the interval contains zero, this
indicates that the null hypothesis $H_0 : \beta_{Area} = 0$ would not be rejected at the 5%
level. We can see from the summary that the p-value is 29.6% — greater than 5%
— confirming this point. Indeed, any point null hypothesis lying within the interval
would not be rejected.

The CI for $\beta_{Adjacent}$ is:

```
lmod.params[5] + lmod.bse[5]*qt
array([-0.06883,  0.06746])
```

Because zero is not in this interval, the null is rejected. Nevertheless, this CI is
relatively wide in the sense that the upper limit is about three times larger than the
lower limit. This means that we are not really that confident about what the exact
effect of the area of the adjacent island on the number of species really is, even
though the statistical significance means we are confident it is negative. A convenient
way to obtain all the univariate intervals is:

```
lmod.conf_int()
                 0          1
Intercept -32.464101  46.600542
Area        -0.070216   0.022339
Elevation    0.208710   0.430219
Nearest     -2.166486   2.184774
Scruz       -0.685093   0.204044
Adjacent    -0.111336  -0.038273
```

The advantage of the confidence interval relative to the corresponding hypothesis
test is that we get information about plausible ranges for the parameters. This is
particularly valuable when the parameter is directly interpretable (for example, as
the difference between two treatments).

The selection of a particular level of confidence level, say 95%, means we can
only make tests at the 5% level. The hypothesis test approach does give a p-value
which allows us to see how the acceptance or rejection of the null hypothesis depends
on the choice of level. Even so, it is dangerous to read too much into the relative
sizes of p-values in determining the practical importance of a predictor because there
is a temptation to view small p-values as indicating an important (rather than just

statistically significant) effect. Confidence intervals are better in this respect because they tell us about the size of the effect.

3.6   Bootstrap Confidence Intervals

The F-based and t-based confidence regions and intervals we have described depend on the assumption of normality. The bootstrap method provides a way to construct confidence statements without this assumption.

To understand how this method works, think about how we might determine the distribution of an estimator without using the mathematical methods of distribution theory. We could repeatedly generate artificial data from the true model, compute the estimate each time and gather the results to study the distribution. This technique, called simulation, is not available to us for real data, because we do not know the true model. Nevertheless, it will reveal the path to a practical solution.

Simulation

The idea is to sample from the known distribution and compute the estimate, repeating many times to find as good an estimate of the sampling distribution of the estimator as we need. For the regression problem, it is easiest to start with a sample from the error distribution since these are assumed to be independent and identically distributed:

1. Generate ε from the known error distribution.
2. Form $y = X\beta + \varepsilon$ from the known β and fixed X.
3. Compute $\hat{\beta}$.

We repeat these three steps many times. We can estimate the sampling distribution of $\hat{\beta}$ using the empirical distribution of the generated $\hat{\beta}$, which we can estimate as accurately as we please by simply running the simulation long enough. This technique is useful for a theoretical investigation of the properties of a proposed new estimator. We can see how its performance compares to other estimators. However, it is of no value for the actual data since we do not know the true error distribution and we do not know the true β.

Bootstrap

The bootstrap emulates the simulation procedure above except instead of sampling from the true model, it samples from the observed data. Remarkably, this technique is often effective. It sidesteps the need for theoretical calculations that may be extremely difficult or even impossible. See Efron and Tibshirani (1993) for a book-length treatment of the topic. To see how the bootstrap method compares with simulation, we spell out the steps involved. In both cases, we consider X fixed.

The bootstrap method mirrors the simulation method, but uses quantities we do know. Instead of sampling from the population distribution, which we do not know in practice, we resample from the data:

1. Generate ε^* by sampling with replacement from $\hat{\varepsilon}_1, \ldots, \hat{\varepsilon}_n$.
2. Form $y^* = X\hat{\beta} + \varepsilon^*$.

3. Compute $\hat{\beta}^*$ from (X, y^*).

This time, we use only quantities that we know. For very small n, it is possible to compute $\hat{\beta}^*$ for every possible sample from $\hat{\varepsilon}_1, \ldots, \hat{\varepsilon}_n$, but usually we can only take as many samples as we have computing power available. This number of bootstrap samples can be as small as 50 if all we want is an estimate of the variance of our estimates but needs to be larger if confidence intervals are wanted.

To implement this, we need to be able to take a sample of residuals with replacement. `np.random.choice()` is good for generating random samples of indices:

```
np.random.choice(np.arange(10),10)
 array([7, 0, 2, 7, 6, 3, 6, 9, 2, 0])
```

You will likely get a different result because the outcome is, by definition, random. To make sure we both get the same result, we have set the random number generator seed using `set.seed`. Here we have used 4000 replications which should take less than a minute to compute. We set up a matrix `coefmat` to store the results. We need the residuals and fitted values from the model which we save as `resids` and `preds`. We repeatedly generate bootstrapped responses as `booty`. The `update` function is more efficient since we are only changing the response, not the predictors. The X-part of the computation does not need to be repeated every time.

```
np.random.seed(123)
breps = 4000
coefmat = np.empty((breps,6))
resids = lmod.resid
preds = lmod.predict()
for i in range(0,breps):
    galapagos['ysamp'] = preds + np.random.choice(resids,30)
    lmodi = smf.ols('ysamp ~ Area + Elevation + \
        Nearest + Scruz  + Adjacent', galapagos).fit()
    coefmat[i,:] = lmodi.params
coefmat = pd.DataFrame(coefmat, columns=("intercept",
    "area","elevation","nearest","Scruz","adjacent"))
coefmat.quantile((0.025,0.975))
```

```
        intercept       area   elevation    nearest     Scruz   adjacent
0.025  -25.038904  -0.060947    0.229760  -1.695215  -0.60020  -0.104559
0.975   41.086507   0.019573    0.416171   2.073810   0.17929  -0.041675
```

The results are saved and formed into a data frame with named columns. We then compute the empirical 2.5% and 97.5% percentiles of the bootstrapped regression coefficients. These form the 95% bootstrap confidence intervals for β. Compare these to those computed using normal theory earlier in the chapter. The position of zero, inside or outside the interval, is the same for both methods so, qualitatively, the results are similar even though the numerical values differ somewhat.

We can plot the estimated density for the `area` coefficients along with the confidence interval as seen in Figure 3.2.

```
coefmat.area.plot.density()
xci = coefmat.area.quantile((0.025,0.975)).ravel()
plt.axvline(x=xci[0], linestyle='--')
plt.axvline(x=xci[1], linestyle='--')
```

We see in both cases that the bootstrap density is symmetrical and similar to a normal density. This is not always so and our method for computing the bootstrap confidence intervals does not rely on symmetry.

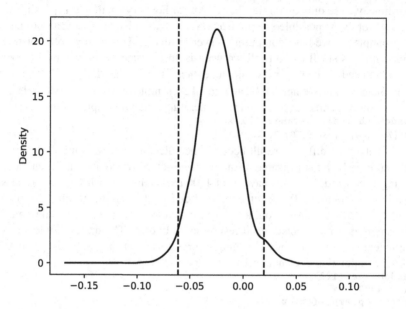

Figure 3.2   Bootstrap regression coefficient density for β_{Area}; 95% confidence interval is shown as dotted lines.

Bootstrap methods can also be used for hypothesis testing although permutation tests are generally better for this purpose. There are alternative resampling methods. We can resample (X, Y) pairs rather than residuals although this seems less attractive, particularly when X is regarded as fixed, such as in designed experiments. There are also more sophisticated methods for constructing the confidence intervals. For more on this, see Efron and Tibshirani (1993) or Davison and Hinkley (1997).

Exercises

1. The dataset `prostate` comes from a study on 97 men with prostate cancer who were due to receive a radical prostatectomy.

   (a) Fit a model with `lpsa` as the response and the other variables as predictors. Show the output.

   (b) Compute 90 and 95% CIs for the parameter associated with `age`. Using just these intervals, what could we have deduced about the p-value for `age` in the regression summary?

   (c) In the text, we made a permutation test corresponding to the F-test for the significance of all the predictors. Execute the permutation test corresponding to the t-test for age in this model.

    (d) Remove all the predictors that are not significant at the 5% level. Test this model against the original model. Which model is preferred?

2. Thirty samples of cheddar cheese were analyzed for their content of acetic acid, hydrogen sulfide and lactic acid. Each sample was tasted and scored by a panel of judges and the average taste score produced. Use the cheddar data to answer the following:

    (a) Fit a regression model with taste as the response and the three chemical contents as predictors. Identify the predictors that are statistically significant at the 5% level.

    (b) Acetic and H2S are measured on a log scale. Fit a linear model where all three predictors are measured on their original scale. Identify the predictors that are statistically significant at the 5% level for this model.

    (c) Can we use an F-test to compare these two models? Explain. Which model provides a better fit to the data? Explain your reasoning.

    (d) If H2S is increased 0.01 for the model used in (a), what change in the taste would be expected?

    (e) What is the percentage change in H2S on the original scale corresponding to an additive increase of 0.01 on the (natural) log scale?

3. Using the teengamb data, fit a model with gamble as the response and the other variables as predictors. The dataset teengamb concerns a study of teenage gambling in Britain. Treat gambling as the response and the sex, status, income and verbal score as predictors.

    (a) Fit the model and display the output.

    (b) Which variables are statistically significant at the 5% level?

    (c) What interpretation should be given to the coefficient for sex?

    (d) Compute the 95% confidence intervals. How is it possible to answer (a) with just this information?

    (e) Fit a model with just income as a predictor and use an F-test to compare it to the full model.

    (f) Test whether status and verbal may be dropped from the full model.

4. Using the sat data:

    (a) Fit a model with total sat score as the response and expend, ratio and salary as predictors. Test the hypothesis that $\beta_{salary} = 0$. Test the hypothesis that $\beta_{salary} = \beta_{ratio} = \beta_{expend} = 0$. Do any of these predictors have an effect on the response?

    (b) Now add takers to the model. Test the hypothesis that $\beta_{takers} = 0$. Compare this model to the previous one using an F-test. Demonstrate that the F-test and t-test here are equivalent.

5. Find a formula relating R^2 and the F-test for the regression.

6. Thirty-nine MBA students were asked about happiness and how this related to their income and social life. The data are found in mba.

(a) Fit a regression model with happy as the response and the other four variables as predictors. Which predictors were statistically significant at the 1% level?

(b) Report on the frequency of the response values using the {value_counts()} function from pandas. What assumption used to perform the t-tests seems questionable in light of this summary?

(c) Use the permutation procedure described in Section 3.3 to test the significance of the money predictor.

(d) Plot a histogram of the permutation t-statistics. Make sure you use the probability rather than frequency version of the histogram.

(e) Overlay an appropriate t-density over the histogram. Hint: create a grid of values using np.linspace(), compute the t-density using sp.stats.t.pdf().

(f) Use the bootstrap procedure from Section 3.6 to compute 90% and 95% confidence intervals for β_{money}. Does zero fall within these confidence intervals? Are these results consistent with previous tests?

7. In the punting data, we find the average distance punted and hang times of 10 punts of an American football as related to various measures of leg strength for 13 volunteers.

(a) Fit a regression model with Distance as the response and the right and left leg strengths and flexibilities as predictors. Which predictors are significant at the 5% level?

(b) Use an F-test to determine whether collectively these four predictors have a relationship to the response.

(c) Relative to the model in (a), test whether the right and left leg strengths have the same effect.

(d) Fit a model to test the hypothesis that it is total leg strength defined by adding the right and left leg strengths that is sufficient to predict the response in comparison to using individual left and right leg strengths.

(e) Relative to the model in (a), test whether the right and left leg flexibilities have the same effect.

(f) Test for left–right symmetry by performing the tests in (c) and (f) simultaneously.

(g) Fit a model where both the strength and flexibility are each summed right plus left. What does the model say about the predictive value of strength and flexibility?

(h) Fit a model with Hang as the response and the same four predictors. Can we make a test to compare this model to that used in (a)? Explain.

Chapter 4

Prediction

Prediction is one of the main uses for regression models. We use subject matter knowledge and the data to build a model $y = X\beta + \varepsilon$. We may have transformed some of the variables or omitted some data points but suppose we have settled on a particular form for the model. Given a new set of predictors, x_0, the predicted response is:

$$\hat{y}_0 = x_0^T \hat{\beta}$$

However, we need to assess the uncertainty in this prediction. Decision makers need more than just a point estimate to make rational choices. If the prediction has a wide CI, we need to allow for outcomes far from the point estimate. For example, suppose we need to predict the high water mark of a river. We may need to construct barriers high enough to withstand floods much higher than the predicted maximum. Financial projections are not so useful without a realistic estimate of uncertainty because we need to make sensible plans for when events do not turn out as well as we predict.

4.1 Confidence Intervals for Predictions

There are two kinds of predictions made from regression models. One is a predicted mean response and the other is a prediction of a future observation. To make the distinction clear, suppose we have built a regression model that predicts the rental price of houses in a given area based on predictors such as the number of bedrooms and closeness to a major highway. There are two kinds of predictions that can be made for a given x_0:

1. Suppose a specific house comes on the market with characteristics x_0. Its rental price will be $x_0^T \beta + \varepsilon$. Since $E\varepsilon = 0$, the predicted price is $x_0^T \hat{\beta}$, but in assessing the variance of this prediction, we must include the variance of ε.

2. Suppose we ask the question — "What would a house with characteristics x_0 rent for on average?". This selling price is $x_0^T \beta$ and is again predicted by $x_0^T \hat{\beta}$ but now only the variance in $\hat{\beta}$ needs to be taken into account.

Most times, we will want the first case, which is called "prediction of a future value," while the second case, called "prediction of the mean response" is less commonly required. We have:

$$\text{var} (x_0^T \hat{\beta}) = x_0^T (X^T X)^{-1} x_0 \sigma^2$$

A future observation is predicted to be $x_0^T \hat{\beta} + \varepsilon$. We do not know the future ε but we expect it has mean zero so the point prediction is $x_0^T \hat{\beta}$. It is usually reasonable to assume that the future ε is independent of $\hat{\beta}$. So a $100(1-\alpha)$ % CI for a single future response is:

$$\hat{y}_0 \pm t_{n-p}^{(\alpha/2)} \hat{\sigma} \sqrt{1 + x_0^T (X^T X)^{-1} x_0}$$

There is a conceptual difference here because previous confidence intervals have been for parameters. Parameters are considered to be fixed but unknown — they are not random under the Frequentist approach we are using here. However, a future observation is a random variable. For this reason, it is better to call this a *prediction interval*. We are saying there is a 95% chance that the future value falls within this interval, whereas it would be incorrect to say that for a parameter.

The CI for the mean response for given x_0 is:

$$\hat{y}_0 \pm t_{n-p}^{(\alpha/2)} \hat{\sigma} \sqrt{x_0^T (X^T X)^{-1} x_0}$$

This CI is typically much narrower than the prediction interval. Although we would like to have a narrower prediction interval, we should not make the mistake of using this version when forming intervals for predicted values.

4.2 Predicting Body Fat

Measuring body fat is not simple. Muscle and bone are denser than fat so an estimate of body density can be used to estimate the proportion of fat in the body. Measuring someone's weight is easy but volume is more difficult. One method requires submerging the body underwater in a tank and measuring the increase in the water level. Most people would prefer not to be submerged underwater to get a measure of body fat so we would like to have an easier method. In order to develop such a method, researchers recorded age, weight, height, and 10 body circumference measurements for 252 men. Each man's percentage of body fat was accurately estimated by an underwater weighing technique. Can we predict body fat using just the easy-to-record measurements?

We load the packages for this chapter:

```
import pandas as pd
import numpy as np
import matplotlib.pyplot as plt
import scipy as sp
import statsmodels.api as sm
import statsmodels.formula.api as smf
import faraway.utils
```

We load in the data and fit a model using all 13 predictors.

```
import faraway.datasets.fat
fat = faraway.datasets.fat.load()
```

We use brozek as the response (Brozek's equation estimates percent body fat from density). Normally, we would start with an exploratory analysis of the data and a detailed consideration of what model to use, but let's be rash and just fit a model and start predicting.

Let's consider the typical man, exemplified by the median value of all the predictors. It's convenient to first create a column for the intercept:

```
fat.insert(0,'Intercept',1)
x0 = fat.iloc[:,np.r_[0,4:7,9:19]].median()
pd.DataFrame(x0).T
```

The final line is simply to print the vector horizontally:

Intercept	age	weight	height	neck	chest
1.0	43.0	176.5	70.0	38.0	99.65

abdom	hip	thigh	knee
90.95	99.3	59.0	38.5

ankle	biceps	forearm	wrist
22.8	32.05	28.7	18.3

Fit the model and make the prediction:

```
lmod = smf.ols('brozek ~ age + weight + height + neck + \
    chest + abdom + hip + thigh + knee + ankle + biceps + \
    forearm + wrist', fat).fit()
x0 @ lmod.params
17.493
```

The predicted body fat for this "typical" man is 17.5%. The same result may be obtained more directly using the `predict` function:

```
lmod.predict(x0)
```

Now if we want a 95% CI for the prediction, we must decide whether we are predicting the body fat for one particular man or the mean body fat for all men have these same characteristics. Here are the two intervals:

```
lmod.get_prediction(x0).summary_frame()
```

	mean	mean_se	mean_ci_lower	mean_ci_upper	obs_ci_lower	obs_ci_upper
0	17.49322	0.278665	16.944255	18.042185	9.61783	25.36861

The prediction interval (marked with obs) ranges from 9.6% body fat up to 25.4%. This is a wide interval since there is a large practical difference between these two limits. One might question the value of such a model. The model has an R^2 of 0.75, but perhaps it is not sufficient for practical use. The confidence interval for the mean response is much narrower, indicating we can be quite sure about the average body fat of the man with the median characteristics. Such information might be useful from a public health perspective where we are concerned about populations rather than individuals.

Extrapolation occurs when we try to predict the response for values of the predictor which lie outside the range of the original data. There are two different types of extrapolation — quantitative and qualitative. Quantitative extrapolation concerns x_0 that are far from the original data. Prediction intervals become wider as we move further from the original data. For multivariate x_0, the concept of distance from the original data is harder to define. In higher dimensions, most predictions will be substantial extrapolations. Let's see what happens with a prediction for values at the 95th percentile of the data:

```
x1 = fat.iloc[:,np.r_[0,4:7,9:19]].quantile(0.95)
pd.DataFrame(x1).T
```

	age	weight	height	neck	chest	abdom	hip	thigh	knee
0.95	67.0	225.65	74.5	41.845	116.34	110.76	112.125	68.545	42.645

	ankle	biceps	forearm	wrist
0.95	25.445	37.2	31.745	19.8

```
lmod.get_prediction(x1).summary_frame()
```

	mean	mean_se	mean_ci_lower	mean_ci_upper	obs_ci_lower	obs_ci_upper
0	30.018044	0.988499	28.07072	31.965369	21.924066	38.112023

We can see that the confidence interval for the mean response is now almost 4% wide compared with the just over 1% width seen in the middle of the data. This is a considerable increase in uncertainty. The prediction interval is only slightly wider because this interval is dominated by the new error ε rather than the uncertainty in the estimation of β.

Unfortunately, there is a much greater source of variation here. We do not know the correct model for this data. We may do our best with the available background information and exploration of the data to find a good model. But no matter how diligent we were, there would still be substantial *model uncertainty*. We have substantial uncertainty about what form the model should take. We can account for *parametric uncertainty* using the methods we have described, but *model uncertainty* is much harder to quantify.

4.3   Autoregression

Consider the data shown in the first panel of Figure 4.1 which shows the monthly number of airline passengers from the early years of air travel. This is an example of a *time series* where a response variable is followed over time. We may want to predict future values of the time series. Linear modeling can be used for this purpose.

We can see an increasing trend which looks more than linear. We can also see some seasonal variation which is increasing over time. For these reasons, we will model the logged passenger numbers. As a first attempt, we might fit a linear function of time to predict this response:

```
import faraway.datasets.air
air = faraway.datasets.air.load()
plt.plot(air['year'], air['pass'])
X = pd.DataFrame({'Intercept':1, 'year':air['year']})
y = np.log(air['pass'])
lmod = sm.OLS(y,X).fit()
plt.plot(air['year'],np.exp(lmod.predict()))
```

We can see that fit captures the general upward trend in numbers but does nothing to model the seasonal variation. Suppose we wish to predict passenger numbers for the next month. We might expect this to depend on the current month. The seasonal variation suggests we also use the observed numbers from 12 months ago and since we are already considering the monthly change from the current month, it makes sense to look at this change from a year ago. Hence we use the numbers from 13 months ago as well. We combine these in a regression model:

$$y_t = \beta_0 + \beta_1 y_{t-1} + \beta_{12} y_{t-12} + \beta_{13} y_{t-13} + \varepsilon_t$$

This is an example of an *autoregressive* process. The response depends on past values of the response. The y_{t-i} are called *lagged* variables. We construct a matrix of lagged variables. We need the current response plus up to 13 lagged variables:

```
air['lag1'] = np.log(air['pass']).shift(1)
air['lag12'] = np.log(air['pass']).shift(12)
air['lag13'] = np.log(air['pass']).shift(13)
airlag = air.dropna()
```

Now we fit the model:

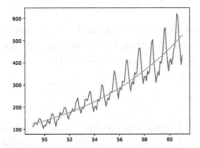 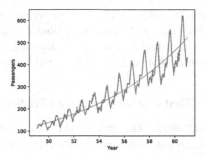

Figure 4.1   Airline passenger numbers 1949 through 1961. An exponential fit is shown on the left. On the right, the fit of an autoregression is shown using dashed lines.

```
X = airlag.loc[:,('lag1','lag12','lag13')]
X.insert(0,'Intercept',1)
y = np.log(airlag['pass'])
lmod = sm.OLS(y,X).fit()
lmod.sumary()
        coefs stderr tvalues pvalues
const   0.138  0.054    2.58  0.0109
lag1    0.692  0.062   11.19  0.0000
lag12   0.922  0.035   26.53  0.0000
lag13  -0.632  0.068   -9.34  0.0000

n=131 p=4 Residual SD=0.042 R-squared=0.99
```

We see all three lagged variables are strongly significant and the fit judged from the R^2 is strong. It is redundant to also include a year term in the model as this linear change over time is now modeled using the lag one variable. We check the fit to the data and observe it in the second panel of Figure 4.1.

```
plt.plot(air['year'], air['pass'])
plt.xlabel("Year")
plt.ylabel("Passengers")
plt.plot(airlag['year'],np.exp(lmod.predict()),linestyle='dashed')
```

Now suppose we want to predict future values. We will need the relevant three observations representing lags 1, 12 and 13:

```
z = np.log(air['pass'].iloc[[-1,-12,-13]]).values
z
```

```
array([6.06842559, 6.03308622, 6.00388707])
```

The current response becomes the lag one value and the other two values also shift. A 95% prediction interval for the (logged) number of passengers is:

```
x0 = pd.DataFrame([{"const":1,"lag1": z[0], "lag12": z[1],
    "lag13": z[2]}])
lmod.get_prediction(x0).summary_frame()
```

	mean	mean_se	mean_ci_lower	mean_ci_upper	obs_ci_lower	obs_ci_upper
0	6.103985	0.006375	6.09137	6.116601	6.020619	6.187351

Now if we want to predict the next month, we will need to plug in the predicted value of 6.104 along with two other shifted lag values. Subsequent predictions will also need to be computed in this recursive manner.

The validity of future predictions depends on the assumption that future data will follow this model. There is no statistical way to check this, so we must accept the possibility that our predictions could be badly wrong and plan accordingly.

4.4   What Can Go Wrong with Predictions?

1. *Bad model.* The statistician does a poor job of modeling the data. In Chapter 6, we present techniques for checking models. These methods can be used to avoid poor model choices. It is also necessary to use information beyond the data about the situation to construct effective models. Sometimes, there is insufficient information in the data itself and other sources must be used.

2. *Bad data.* A model is only good as the data used to construct it. In Section 3.4, we discussed various mechanisms under which data might be obtained. These will have consequences for the quality and versatility of the model. Sometimes, most of the data may be good, but a few bad cases can have serious consequences. In Section 6.2 we present methods for detecting these bad cases and in Section 8.4, we discuss model fitting methods that are robust to these bad cases.

3. *Quantitative extrapolation.* This happens when we try to predict outcomes for cases with predictor values much different from what we saw in the data. An example is seen above. This is a practical problem in assessing the risk from low exposure to substances which are dangerous in high quantities — consider second-hand tobacco smoke, asbestos and radon. It is more likely to be a problem for models with more predictors as it is harder to determine whether a combination of predictors is unusual.

   One advantage of linear models is they behave in quite predictable ways depending on the input predictor values. For more complex non-linear models, predictions can be surprising.

4. *Qualitative extrapolation.* We try to predict outcomes for observations that come from a different population. For example, suppose we used the models above to predict body fat for women? This is a common problem because circumstances are always changing and it's hard to judge whether the new case is comparable. In this case, we might judge that the model would not work for women, but it would work for men from another country.

   We prefer experimental data to observational data, but sometimes experience from the laboratory does not transfer to real life.

5. *Overconfidence* due to overfitting. Prediction models are often dissappointing in that the future observed values lie outside the prediction intervals more often than one might hope. A common cause is that modelers tend to fit the data they have far too well. They attribute too much to systematic effects and too little to random variation. This can lead to unrealistically small $\hat{\sigma}$. We defer further discussion of this until Chapter 10 when we have more models to choose from.

6. *Black swans.* Sometimes errors can appear to be normally distributed because

you haven't seen enough data to be aware of extremes. This is of particular concern in financial applications where stock prices are characterized by mostly small changes (normally distributed) but with infrequent large changes (usually falls). It is difficult to accommodate these possibilities within the model since we have insufficient information. It is wise for the users of prediction models to consider the consequences of the unexpected in more qualitative ways.

Exercises

1. For the `prostate` data, fit a model with `lpsa` as the response and the other variables as predictors.

   (a) Suppose a new patient with the following values arrives:

```
lcavol  lweight      age     lbph      svi      lcp
1.44692  3.62301 65.00000  0.30010  0.00000 -0.79851
gleason     pgg45
7.00000 15.00000
```

   Predict the `lpsa` for this patient along with an appropriate 95% uncertainty interval.

   (b) Repeat the last question for a patient with the same values except that he is age 20. Explain why the interval is wider.

   (c) For the model of the previous question, remove all the predictors that are not significant at the 5% level. Now recompute the predictions of the previous question. Are the intervals wider or narrower? Which predictions would you prefer? Explain.

2. Using the `teengamb` data, fit a model with `gamble` as the response and the other variables as predictors.

   (a) Predict the amount that an average male with average (given these data) status, income and verbal score would gamble along with an appropriate 95% interval.

   (b) Repeat the prediction for males with maximal values (for this data) of status, income and verbal score. Which interval is wider and why is this result expected?

   (c) Fit a model with `sqrt(gamble)` as the response but with the same predictors. Now predict the response and give a 95% prediction interval for the individual in (a). Take care to give your answer in the original units of the response.

   (d) Repeat the prediction for the model in (c) for a female with `status=20`, `income=1`, `verbal = 10`. What is wrong with the prediction interval?

3. The `snail` dataset contains percentage water content of the tissues of snails grown under three different levels of relative humidity and two different temperatures.

   (a) Use the command `xtabs(water ~ temp + humid, snail)/4` to produce a table of mean water content for each combination of temperature and humidity. Can you use this table to predict the water content for a temperature of 25°C and a humidity of 60%? Explain.

(b) Fit a regression model with the water content as the response and the temperature and humidity as predictors. Use this model to predict the water content for a temperature of 25°C and a humidity of 60%?

(c) Use this model to predict water content for a temperature of 30°C and a humidity of 75%? Compare your prediction to the prediction from (a). Discuss the relative merits of these two predictions.

(d) The intercept in your model is 52.6%. Give two values of the predictors for which this represents the predicted response. Is your answer unique? Do you think this represents a reasonable prediction?

(e) For a temperature of 25°C, what value of humidity would give a predicted response of 80% water content.

4. We can obtain the number of monthly deaths from lung diseases for people in the UK from 1974 to 1979 with
```
import statsmodels.api as sm
deaths = sm.datasets.get_rdataset("deaths","MASS").data
deaths.head()
```

(a) Make an appropriate plot of the data. At what time of year are deaths most likely to occur?

(b) Fit an autoregressive model of the same form used for the airline data. Are all the predictors statistically significant?

(c) Use the model to predict the number of deaths in January 1980 along with a 95% prediction interval.

(d) Use your answer from the previous question to compute a prediction and interval for February 1980.

(e) Compute the fitted values. Plot these against the observed values. Note that you will need to select the appropriate observed values. Do you think the accuracy of predictions will be the same for all months of the year?

5. For the fat data used in this chapter, a smaller model using only age, weight, height and abdom was proposed on the grounds that these predictors are either known by the individual or easily measured.

(a) Compare this model to the full 13 predictor model used earlier in the chapter. Is it justifiable to use the smaller model?

(b) Compute a 95% prediction interval for median predictor values and compare the results to the interval for the full model. Do the intervals differ by a practically important amount?

(c) For the smaller model, examine all the observations from case numbers 25 to 50. Which two observations seem particularly anomalous?

(d) Recompute the 95% prediction interval for median predictor values after these two anomalous cases have been excluded from the data. Did this make much difference to the outcome?

Chapter 5

Explanation

Linear models can be used for prediction or explanation. Prediction is not simple but it is conceptually easier than explanation. We have been deliberately vague about the meaning of explanation. Sometimes explanation means causation but sometimes it is just a description of the relationships between the variables. Causal conclusions require stronger assumptions than those used for predictive models. This chapter looks at the conditions necessary to conclude a causal relationship and what can be said when we lack these conditions.

5.1 Simple Meaning

Start by loading the packages used in this chapter:
```
import pandas as pd
import numpy as np
import matplotlib.pyplot as plt
import scipy as sp
import statsmodels.api as sm
import statsmodels.formula.api as smf
import faraway.utils
```
Let's consider the Galápagos Islands example:
```
import faraway.datasets.galapagos
galapagos = faraway.datasets.galapagos.load()
lmod = smf.ols(
    'Species ~ Area + Elevation + Nearest + Scruz   + Adjacent',
    galapagos).fit()
lmod.sumary()
```
	coefs	stderr	tvalues	pvalues
Intercept	7.068	19.154	0.37	0.7154
Area	-0.024	0.022	-1.07	0.2963
Elevation	0.319	0.054	5.95	0.0000
Nearest	0.009	1.054	0.01	0.9932
Scruz	-0.241	0.215	-1.12	0.2752
Adjacent	-0.075	0.018	-4.23	0.0003

```
n=30 p=6 Residual SD=60.975 R-squared=0.77
```
What is the meaning of the coefficient for Elevation $\hat{\beta} = 0.319$?

In a few examples, mostly from the physical sciences or engineering, β might represent a real physical constant. For example, we might attach weights to a spring and measure the extension. Here $\hat{\beta}_1$ will estimate a physical property of the spring. In such examples, the model is a representation of a physical law. But for the Galápagos data, our model has no such strong theoretical underpinning. It is an empirical model that we hope is a good approximation to reality. $\hat{\beta}_1$ has no direct physical meaning in this example.

Let's start with a simple interpretation: *a unit increase in x_1 will produce a change of $\hat{\beta}_1$ in the response y.* This statement is rather imprecise regarding the nature of the relationship between x_1 and y, so let's be more specific: Compare two islands where the second island has an elevation one meter higher than the first. We predict that the second island will have 0.32 species more than the first. Let's set aside the fact that the number species can only take integer values because the model is only intended as an approximation. We could specify that the second island was 100m higher than the first resulting in a predicted difference of about 32 species. There are more serious problems with this interpretation.

Consider an alternative model for the number of species that uses only the elevation variable:

```
lmodr = smf.ols('Species~Elevation',galapagos).fit()
lmodr.sumary()
```

```
          coefs stderr tvalues pvalues
Intercept 11.335 19.205    0.59  0.5598
Elevation  0.201  0.035    5.80  0.0000
```

```
n=30 p=2 Residual SD=78.662 R-squared=0.55
```

We can see that the predicted difference for an increase of 100m in elevation is now about 20 species more, in contrast to the 32 found for the larger model. The different values illustrate why we cannot interpret a regression coefficient for a given predictor without reference to the other predictors included in the model. We show the two fits for the elevation in Figure 5.1. To show the relationship between elevation and the response for the full five-predictor model, we fix the other predictors at some typical values. We have chosen the means. We then compute the predicted response for the minimum and maximum observed values of elevation. A line connecting these points will represent the fit as elevation varies:

```
x0 = galapagos.mean()
xdf = pd.concat([x0,x0],axis=1).T
xrange = [np.min(galapagos.Elevation),np.max(galapagos.Elevation)]
xdf['Elevation'] = xrange
```

Now we make the plot:

```
plt.scatter(galapagos.Elevation, galapagos.Species)
plt.xlabel("Elevation")
plt.ylabel("Species")
plt.plot(xrange,lmodr.predict(xdf),"-")
plt.plot(xrange,lmod.predict(xdf),"--")
```

This is called an *effect plot.* It is a good way of visualizing the meaning of the model for a given predictor. We can even compare the effects to other models as we have done in the example.

We need to be more specific in our interpretation which should now read: *a unit increase in x_1 with the other (named) predictors held constant will produce a change of $\hat{\beta}_1$ in the response y.* For the simple one-predictor model, a change in elevation will also be associated with a change in the other predictors. This explains why the two predictions are different.

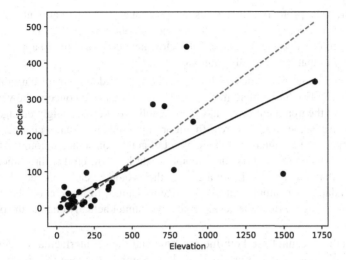

Figure 5.1   The fit for the simple model with just elevation as a predictor is shown as the solid line. The predicted response as elevation varies and the other four predictors held at their mean values is shown as a dashed line.

Unfortunately, our revised interpretation merely brings to light further difficulties. The idea of holding variables constant makes no sense for observational data such as in our Galápagos example. These observables are not under our control. We cannot change them except by some fantastic feat of engineering. There are also likely to be other variables which we have not measured and may not be aware of that also have some connection to the species diversity. We cannot possibly hold these constant.

Furthermore, our explanation contains no notion of causation. We can predict that taller islands have greater species diversity, but we should not say altitude causes it. In the next section, we look at how causality might be established. For now we are able to make predictions and compare these predictions to develop some insights. The comparisons give some meaning to the regression coefficients but the information is conditional and tentative.

5.2   Causality

The meaning of causality has occupied philosophers for centuries. We take the view that the causal effect of an action is the difference between the outcomes where the action was or was not taken. For example, suppose a study compares a treatment with a control as applied to a set of patients. Let $T = 0$ for the control and $T = 1$ for the treatment. Now let y_i^T be the response for patient i when T applies. The causal effect for patient i is then defined as

$$\delta_i = y_i^1 - y_i^0$$

Now the practical problem is that we usually cannot apply both treatment and control at the same time. We only get to see one of the two outcomes: (y_i^0, y_i^1). The outcome we do not see is called the *counterfactual*. How near can we come to approximating this quantity δ_i that we normally cannot observe?

There are some difficulties with the counterfactual definition of causality. For some variables, like a treatment in a clinical trial, it is easy to conceive how a different version of the treatment might have been applied or how we might change future treatments. But some variables are difficult or impossible to actually change. For example, suppose we are interested in the effect of gender on some outcome. Although it seems reasonable to ask how an outcome might have differed if the subject were male rather than female, this is not a change that would be easy to make. Similarly, with the Galápagos example, although it seems natural to think about how physical geography might affect species diversity, we cannot actually change the physical geography.

Although we cannot see both the outcome and the counterfactual outcome, we may come close in some circumstances. For example, suppose the treatment is a skin cream. We could apply the treatment to one side of the face and leave the other as a control. Or we could apply the control first (which might be no treatment) and then apply the cream later. We could then observe the difference in the post- and pre-treatment responses. However, even in such cases, it is easy to see that some assumptions will be necessary about how local the effect of the treatment is in time and space. We will also need to be wary of external variables that might have differential effects. If these assumptions are reasonable, then we may proceed but in many other circumstances we will have doubts or it simply may not be possible to apply more than one treatment to a unit.

5.3   Designed Experiments

In a designed experiment, we have control over T. For example, suppose we wish to compare two physical exercise regimes. The *experimental units* are the people we use for the study. There may be some other potential predictors which we can control such as the amount of time spent excercising or the type of equipment used. Some other predictors might not be controlled, but can be measured, such as the physical characteristics of the people. Still other predictors may not be controlled or measured. We may know about these predictors, or we may be unaware of them. Other possible variables, such as the temperature in the room, might be held constant. Our control over the conditions of the experiment allows us to make stronger conclusions from the analysis. Randomization is the key to success.

Consider the easiest case where we will vary only T. We have some number of subjects available for the experiment. Although we would like to know the individual causal effects δ_i, this is not possible because only one level of T can be assigned to a given subject at a given moment in time. However, we can aspire to estimate the average value of δ over the group. We should randomly assign subjects to treatment or control. Typically, we will do this so that there are an equal number of units in the two groups, but this is not essential. There are two compelling reasons to use randomization.

We know that the subjects will vary in ways which may affect the response. Some of these variables may not be obvious or measurable. Randomization is the only reliable way to make sure that the two groups are not unbalanced in some way that favors either the treatment or the control. But this first argument for randomization only ensures that the groups will be balanced on the average. It may well happen that the particular realization of the randomization is unbalanced. We have a second and more compelling argument to use. A permutation test (see Section 3.3) can be used to test the null hypothesis of no difference between the groups. The justification of the permutation test relies on the observed allocation being chosen randomly from all possible allocations to groups. By using randomization, we validate this assumption. It is true we could get an unbalanced allocation by chance but we also understand that we will reject true null hypotheses by chance. So this possibility is built into the reasoning behind the test. Remember that the permutation test tends to agree with the normal-based test provided the linear model assumptions are justified. So it is usually alright if we use the normal-based test rather than actually do the work required for the permutation test.

The results of this test apply only to the subjects used in the experiment unless we can reasonably claim that these subjects are representative of a larger population. It is best if they are a random sample but this is rarely the case in practice, so further assumptions are necessary if we wish to make wider claims.

Now suppose we are aware that the experimental units differ in identifiable ways. For example, some subjects may be male and others female. We may wish to incorporate this into the design by restricting the randomization. The obvious strategy is to randomize separately the assignment of the treatments within males and females. This will ensure that the groups are balanced by sex. In this example, sex is called a *blocking* variable. Such designs are covered in detail in Chapter 17. In other cases, there may be variables which are not properties of the experimental units (like sex) but can be assigned (like time of exercise). In both situations, we will want to arrange the design so that it is *orthogonal*. This concept is explained in Section 2.11. This is not essential for causal conclusions, but it does greatly simplify them.

We have considered a binary treatment variable T, but the same goes for continuous potential causal variables. Some additional assumptions may be necessary regarding the functional form of the relationship — for example, that T has a linear effect on the response.

5.4 Observational Data

Sometimes it is not practical or ethical to collect data from a designed experiment. We cannot control the assignment of T and so we can only obtain observational data. In some situations, we can control which cases we observe from those potentially available. A *sample survey* is used to collect the data. A good survey design can allow stronger and wider conclusions, but the data will still be observational.

On the 8th January 2008, primaries to select US presidential candidates were held in New Hampshire. In the Democratic Party primary, Hillary Clinton defeated Barack Obama contrary to the expectations of pre-election opinion polls. Two different voting technologies were used in New Hampshire. Some wards (administrative districts) used paper ballots, counted by hand, while others used optically scanned ballots, counted by machine. Among the paper ballots, Obama had more votes than Clinton, while Clinton defeated Obama on just the machine-counted ballots. Since the method of voting should make no causal difference to the outcome, suspicions were raised regarding the integrity of the election. The data was derived from Herron et al. (2008) where a more detailed analysis may be found.

```
import faraway.datasets.newhamp
newhamp = faraway.datasets.newhamp.load()
newhamp.groupby('votesys').agg({'Obama': sum, 'Clinton': sum})
        Obama  Clinton
votesys
D       86353   96890
H       16926   14471
```

We focus our interest on the voting system variable votesys which can either be 'D' for digital or 'H' for hand. We use the proportion voting for Obama in each ward as the response. Strictly speaking, this is a binomial response and we should model it accordingly. Nevertheless, the normal is a good approximation for the binomial given a large enough sample and probabilities not close to zero or one. We have that in this sample. Even so, a binomial variance is $np(1-p)$ for proportion p and sample size n. Both of these vary in this example and so the assumption of equal variance is violated. We can fix this problem by using weights as described in Section 8.2, but this would make no appreciable difference to our discussion here, so we ignore it.

Let us fit a linear model with just voting system as a predictor. We create an indicator variable for this treatment where 1 represents hand and 0 represents digital voting:

```
newhamp['trt'] = np.where(newhamp.votesys == 'H',1,0)
lmodu = smf.ols('pObama ~ trt',newhamp).fit()
lmodu.sumary()
          coefs stderr tvalues pvalues
Intercept 0.353  0.005   68.15  0.0000
trt       0.042  0.009    4.99  0.0000

n=276 p=2 Residual SD=0.068 R-squared=0.08
```

The model takes the form:

$$y_i = \beta_0 + \beta_1 T_i + \varepsilon_i$$

When digital voting is used, the predictor is set to 0 and the predicted proportion is therefore $\hat{\beta}_0 = 35\%$. When hand voting is used, the prediction is $\hat{\beta}_1 = 4\%$ higher. We see from the very small p-value of 0.0000011 that this difference is significantly different. So we are quite sure that Obama received a significantly higher proportion of the vote in the hand voting wards. The question is why? Did the voting method have some causal effect on the outcome?

Suppose that the correct model involved some third variable Z and took the form:

$$y_i = \beta_0^* + \beta_1^* T_i + \beta_2^* Z_i + \varepsilon_i$$

and suppose that this Z was linked to T by:

$$Z_i = \gamma_0 + \gamma_1 T_i + \varepsilon'_i$$

Z is sometimes called a *confounding* variable. Substituting the latter into the former, we find the coefficient for T is $\beta_1^* + \beta_2^*\gamma_1$. Compare this to the β_1 in the initial model where we do not include Z. There are two ways in which the conclusion could be the same for the two models. We could have $\beta_2^* = 0$ when Z has no effect on the response or $\gamma_1 = 0$ where the treatment has no effect on Z. Otherwise Z will have an effect on our conclusions and the initial model which excludes Z will provide a biased estimate of the treatment effect. In a designed experiment, we have $\gamma_1 = 0$ by the randomization in the assignment of T.

Does such a third variable Z exist for the New Hampshire voting example? Consider the proportion of votes for Howard Dean, a Democratic candidate in the previous presidential campaign in 2004. We add this term to the model:

```
lmodz = smf.ols('pObama ~ trt+Dean',newhamp).fit()
lmodz.sumary()
          coefs stderr tvalues pvalues
Intercept  0.221  0.011   19.65  0.0000
trt       -0.005  0.008   -0.61  0.5407
Dean       0.523  0.042   12.55  0.0000
```

n=276 p=3 Residual SD=0.054 R-squared=0.42

We see that the effect of the voting system is no longer statistically significant with a p-value of 0.54. The proportion voting for Dean shows a positive relationship to the proportion voting for Obama. We can also see that this third variable is related to our "treatment" variable:

```
lmodc = smf.ols('Dean ~ trt',newhamp).fit()
lmodc.sumary()
          coefs stderr tvalues pvalues
Intercept 0.251  0.006   41.99  0.0000
trt       0.090  0.010    9.18  0.0000
```

n=276 p=2 Residual SD=0.079 R-squared=0.24

We can see that there is an active confounder in this situation. What should we conclude? In the next section, we show how we can use counterfactual notions to clarify the effect of the voting system on preferences for Obama.

5.5 Matching

People vote for candidates based on political and character preferences. The proportion of voters choosing Howard Dean in the previous Democratic primary tells us something about the aggregate preferences of the voters in each ward in the 2008 primary. Suppose that we had been allowed to do an experiment in 2008 and we were permitted to assign the voting systems to the wards. How would we do this?

Consider a clinical trial where we compare a treatment to a control. We have a pool of available subjects who differ in identifiable ways such as sex, age, overall health condition and so on that might affect the response. We could simply randomly divide the pool in two, assigning one group to the treatment and one to the control.

But the random division could be unbalanced with respect to the confounders. Although this possibility is built into the testing procedure, the approach does not make the best use of the available information. Often a better approach is to form matched pairs where the two members of each pair are as alike as possible with respect to the confounders. For example, we would match a healthy older man with another healthy older man. Within the pairs, treatment and control are randomly assigned. We can then determine the effect of the treatment by looking at the difference in the response for each pair. We do this secure in the knowledge that we have adjusted for the effect of confounding by balancing these differences between treatment and control.

Let's return to the New Hampshire primary and apply the same procedure. We would form matched pairs based on similarity of proportion formerly supporting Dean. Within each pair, we would make a random assignment. Unfortunately, in real life, we cannot make the random assignment but we can find pairs of wards with similar values of Dean proportion where one uses hand voting and the other uses digital.

We use a "greedy matching" algorithm to find these pairs. For each point in the smaller hand-voting group, we look for a case in the digital-voting group with a Dean fraction within 0.01. We will not be able to match all cases in the hand-voting group because a close enough match may not exist or because the available cases in the digital group have already been matched. There are better and more sophisticated algorithms but this one is easy to program:

```
sg = newhamp.Dean[newhamp.trt == 1]
bg = newhamp.Dean[newhamp.trt == 0]
ns = len(sg)
mp = np.full([ns,2],-1)
for i in range(ns):
    dist = abs(sg.iloc[i]-bg)
    if(dist.min() < 0.01):
        imin = dist.idxmin()
        mp[i,:] = [sg.index[i], imin]
        bg = bg.drop(index = imin)
mp = mp[mp[:,0] > -1,:]
mp
array([[  3, 212],
       [ 16,  19],
       [ 17,   5],
       [ 18,  90],
etc..
```

We examine the first few pairs of matches and see that first pair uses cases 3 and 212 which we look at specifically. For this pair of treatment and control, we see that the value of Dean is very similar.

```
newhamp.iloc[[76,  84],[0,6,11]]
```

	votesys	Dean	pObama
3	H	0.28495	0.343284
212	D	0.28457	0.382932

We show all the matches in Figure 5.2. Notice that some wards go unmatched because there is not a close match of the other type available.

```
sy = newhamp.pObama[newhamp.trt == 1]
by = newhamp.pObama[newhamp.trt == 0]
```

```
bg = newhamp.Dean[newhamp.trt == 0]
plt.scatter(bg, by, marker="^", s=2)
plt.scatter(sg, sy, marker="o", s=2)
plt.xlabel("Dean proportion")
plt.ylabel("Obama proportion")
for i in range(len(mp)):
    plt.plot([sg.loc[mp[i,0]], bg.loc[mp[i,1]]],
             [sy.loc[mp[i,0]], by.loc[mp[i,1]]])
```

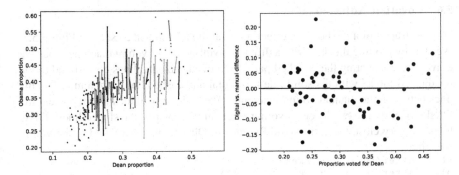

Figure 5.2   On the left, we show the matched pairs hand-voting wards are shown as triangles and digital-voting wards as circles. Matched pairs are linked by a line. On the right, we show the matched pair differences.

We compute the difference between the pairs and perform a one sample t-test.
```
pdiff = sy.loc[mp[:,0]].ravel()-by.loc[mp[:,1]].ravel()
sp.stats.ttest_1samp(pdiff,0)
Ttest_1sampResult(statistic=-1.44, pvalue=0.154)
```
We plotted the differences in the second panel of Figure 5.2.
```
plt.scatter(sg.loc[mp[:,0]], pdiff)
plt.axhline(0)
plt.xlabel('Proportion voted for Dean')
plt.ylabel('Digital vs. manual difference')
```
We see that the differences are not statistically significantly different from zero. This is confirmed by the plot where we can see that the matched pairs show no clear preference for hand or digital voting. Furthermore, the difference doesn't appear to depend on the Dean vote.

We were not able to do an experiment and we certainly were not able to view the counterfactual, but we achieved the next best thing. We have matched wards with similar political outlooks by using the prior Dean vote proportion. One ward has used digital voting and the other has used the manual method. Based on this comparison, we see no significant difference between the two methods. The observed unadjusted difference is because voters inclined to pick Obama are also more likely to be present in hand-voting wards.

Now we haven't disproved any conspiracy theories about fixed votes, but we have come up with a more plausible explanation. We do not claim our new explanation is necessarily true because there may be yet further variables that drive the relationship.

Nevertheless, it is a basic principle of science that we should prefer simpler theories provided they are tenable.

There is much more to matching. We have presented a simple approach but there are better ways to do the analysis and more sophisticated ways to do the matching in the presence of covariates, but now we return to our earlier linear modeling and consider how this relates to the matched analysis.

5.6 Covariate Adjustment

We construct a plot to show the effect of covariate adjustment as seen in Figure 5.3. We start by plotting the data with a different symbol for the voting machine type. We draw lines connecting the matched pairs. The two horizontal lines correspond to the unadjusted model with the solid line for digital and dashed line for manual. Obama does better on the manual voting machines. We add the two lines that represent the fitted model for the two levels of voting mechanism shown as the sloped lines. These lines are very close indicating no difference in Obama support related to the voting machine.

```
plt.scatter(bg, by, marker="^", s=2)
plt.scatter(sg, sy, marker="o", s=2)
plt.xlabel("Dean proportion")
plt.ylabel("Obama proportion")
for i in range(len(mp)):
    plt.plot([sg.loc[mp[i,0]], bg.loc[mp[i,1]]],
             [sy.loc[mp[i,0]], by.loc[mp[i,1]]],color="0.75")
drange = [0.1,0.6]
x0 = pd.DataFrame({"trt": (0,0), "Dean": drange})
x1 = pd.DataFrame({"trt": (1,1), "Dean": drange})
plt.plot(drange,lmodu.predict(x0))
plt.plot(drange,lmodu.predict(x1), linestyle="dashed")
plt.plot(drange,lmodz.predict(x0))
plt.plot(drange,lmodz.predict(x1), linestyle="dashed")
```

If we consider the expected response for a fixed value of Dean, the difference between digital and hand is given by the vertical distance between the two sloped lines. Compare this to the matching approach where we find pairs of wards, one digital, one hand, which have almost the same value of Dean. Each pair represents a local realization of that vertical difference. When we average these pairwise differences, we also estimate the vertical difference. Thus matching on a covariate and fitting a linear model where we adjust for the covariate by including it in the model are two ways to estimate the same thing.

Both approaches to drawing conclusions from observation studies are useful. The covariate adjustment method, sometimes called "controlling for a covariate", is easier to use and extends well to multiple confounders. However, it does require that we specify the functional form of the covariate in the model in an appropriate way. In our example, it is clear from the plots that a linear form is suitable, but in higher dimensions this can be harder to check. The matching approach is more robust in that it does not require we specify this functional form. In our example, the two groups overlapped substantially so that it was possible to obtain a large number of matched pairs. When there is less overlap, the covariate adjustment approach can still make

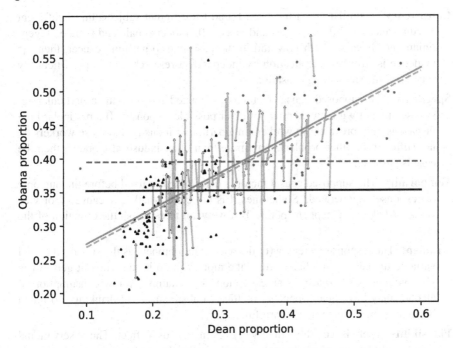

Figure 5.3   Linear models for Obama proportion. The solid lines represent the digital-voting wards while the dashed lines represent the hand-voting wards. The horizontal lines result from an unadjusted model, while the sloped lines result from the model that includes the proportion voting for Dean. Matched pairs are linked by gray lines.

full use of the data, although it is vulnerable to the dangers of extrapolation when making comparisons. In the rest of this book, we shall generally use the covariate adjustment approach, but the matching approach should not be discounted and can provide valuable insights.

5.7   Qualitative Support for Causation

In observational studies, there is a limit to what can be achieved with statistical manipulations. Sir Bradford Hill was a central figure in establishing the causal link between smoking and lung cancer. In Hill (1965) he laid out several general considerations that can reinforce the case for a causal link:

Strength We do not mean a high correlation or a small p-value but that $\hat{\beta}$ is large in practical terms. Known covariates can be adjusted for, while unobserved and unsuspected confounding variables could easily lead to small effects. It is less credible that some variable whose existence was previously unknown could counteract a large effect.

Consistency A similar effect has been found for different subjects under different circumstances at different times and places. If smokers, male and female, Argentinian and Swedish, both now and in the past tend to get lung cancer, then the evidence is stronger. Replication by independent research groups is particularly important in establishing causation.

Specificity The supposed causal factor is associated mostly with a particular response and not with a wide range of other possible responses. If a particular lung disease is only prevalent in workers in a particular industry and those workers do not suffer from other problems any more than other industrial workers, then the case is stronger.

Temporality The supposed causal factor is determined or fixed before the outcome or response is generated. Sometimes it is not clear whether X causes Y or vice versa. It helps if X happens before Y if we want to establish the direction of the effect.

Gradient The response increases (or decreases) monotonely as the supposed causal variable increases. In other words, "the more you do it, the worse it gets." The phenomenom of *hormesis* is an exception. For example, a low consumption of alcohol may have health benefits relative to abstinence, but drinking too much becomes progressively more harmful.

Plausibility There is a credible theory suggesting a causal effect. The observational study might be part of an effort to support a particular theory.

Experiment A natural experiment exists where subjects have apparently been randomly assigned values of the causal variable. For example, US states have different laws that have differential effects on the causal factor. Sometimes it is reasonable to view this as a form of random assignment. Generally, we conduct observational studies because experiments are not practical, but sometimes it is possible to do related experiments. For example, animal experiments might tell us something about effects for humans.

Not all of these might apply, but causation might still be present. All of these might apply, but we may not be sure of causation. Nevertheless, these considerations do add to the weight of evidence even though we cannot express the effect numerically.

Exercises

1. Use the `teengamb` data with `gamble` as the response. We focus on the effect of `sex` on the response and so we include this predictor in all models. There are eight possible models that include all, some, or none of the other three predictors. Fit all these models and report on the coefficient and significance of `sex` in each case. Comment on the stability of the effect.

2. Use the `odor` dataset with odor as the response and `temp` as a predictor. Consider all possible models that also include all, some or none of the other two predictors. Report the coefficient for temperature, its standard error, t-statistic and p-value in

each case. Discuss what stays the same, what changes and why. Which model is best?

3. Use the `teengamb` data for this question.

   (a) Make a plot of `gamble` on `income` using a different plotting symbol depending on the `sex`.

   (b) Fit a regression model with `gamble` as the response and `income` and `sex` as predictors. Display the regression fit on the plot.

   (c) Use the method from the chapter to find matches on `sex`. Use the same parameters as in the chapter. How many matched pairs were found? How many cases were not matched?

   (d) Make a plot showing which pairs were matched.

   (e) Compute the differences in `gamble` for the matched pairs. Is there a significant non-zero difference?

   (f) Plot the difference against `income`. In what proportion of pairs did the female gamble more than the male?

   (g) Do the conclusions from the linear model and the matched pair approach agree?

4. Thirty-nine MBA students were asked about happiness and how this related to their income and social life. The data are found in `happy`.

   (a) Fit a regression model with `happy` as the response and the other four variables as predictors. Give an interpretation for the meaning of the `love` coefficient.

   (b) The `love` predictor takes three possible values but mostly takes the value 2 or 3. Create a new binary predictor called `clove` which takes the value zero if `love` is 2 or less. Use this new predictor to replace `love` in the regression model and interpret the meaning of the corresponding coefficient. Do the results differ much from the previous model?

   (c) Fit a model with only `clove` as a predictor and interpret the coefficient. How do the results compare to the previous outcome?

   (d) Make a plot of `happy` on `work`, distinguishing the value `clove` by using a plotting symbol. Use jittering to distinguish overplotted points.

   (e) Use the command `pd.crosstab(mba.clove,mba.work)` to produce a crosstabulation. If we wanted to match pairs on `clove` with the same value of `work`, what is the maximum number of 1 to 1 matches we could achieve?

   (f) For each value of `work`, compute the mean difference in `happy` for the two levels of `clove`. Compute an average of these differences. Which coefficient computed earlier would be the most appropriate comparison for this average?

Chapter 6

Diagnostics

The estimation of and inference from the regression model depend on several assumptions. These assumptions should be checked using *regression diagnostics* before using the model in earnest. We divide the potential problems into three categories:

Error We have assumed that $\varepsilon \sim N(0, \sigma^2 I)$ or, in words, that the errors are independent, have equal variance and are normally distributed.

Model We have assumed that the structural part of the model, $Ey = X\beta$, is correct.

Unusual observations Sometimes just a few observations do not fit the model. These few observations might change the choice and fit of the model.

Diagnostic techniques can be graphical, which are more flexible but harder to definitively interpret, or numerical, which are narrower in scope, but require less intuition. When searching for a good model, the first one we try may prove to be inadequate. Regression diagnostics often suggest specific improvements, which means model building is an iterative and interactive process. It is quite common to repeat the diagnostics on a succession of models.

6.1 Checking Error Assumptions

We wish to check the independence, constant variance and normality of the errors, ε. The errors are not observable, but we can examine the residuals, $\hat{\varepsilon}$. These are not interchangeable with the error, as they have somewhat different properties. Recall that $\hat{y} = X(X^T X)^{-1} X^T y = Hy$ where H is the hat matrix, so that:

$$\hat{\varepsilon} = y - \hat{y} = (I - H)y = (I - H)X\beta + (I - H)\varepsilon = (I - H)\varepsilon$$

Therefore, var $\hat{\varepsilon} = $ var $(I - H)\varepsilon = (I - H)\sigma^2$ assuming that var $\varepsilon = \sigma^2 I$. We see that although the errors may have equal variance and be uncorrelated, the residuals do not. Fortunately, the impact of this is usually small and diagnostics can reasonably be applied to the residuals in order to check the assumptions on the error.

6.1.1 Constant Variance

It is not possible to check the assumption of constant variance just by examining the residuals alone — some will be large and some will be small, but this proves nothing. We need to check whether the variance in the residuals is related to some other quantity.

The most useful diagnostic is a plot of $\hat{\varepsilon}$ against $\hat{y}$. If all is well, you should see constant symmetrical variation (known as homoscedasticity) in the vertical ($\hat{\varepsilon}$) direction. Nonconstant variance is also called heteroscedasticity. Nonlinearity in the structural part of the model can also be detected in this plot. In Figure 6.1, three distinct cases are illustrated. We have generated these from known models and here is how we did it. First we need to load the packages for this chapter:

```
import pandas as pd
import numpy as np
import matplotlib.pyplot as plt
import scipy as sp
import statsmodels.api as sm
import statsmodels.formula.api as smf
import faraway.utils
import seaborn as sns
```

We start by generating some randomly spaced fitted values:

```
n = 50
np.random.seed (123)
x = np.random.sample(n)
```

We have set the random number generator seed so that if you repeat this, you will get the exact same plots.

Now we generate some normally distributed residuals and make the plot:

```
y = np.random.normal(size=n)
plt.scatter(x,y)
plt.title("No problem plot")
plt.axhline(0)
```

The next example increases the variance by multiplying it by the fitted values:

```
plt.scatter(x,y*x)
plt.title("Increasing variance")
plt.axhline(0)
```

In the final example, we introduce a nonlinearity:

```
y = np.cos(2*x*np.pi) + np.random.normal(size=n)
plt.scatter(x,y)
plt.title("Lack of fit plot")
sx = np.sort(x)
plt.plot(sx,np.cos(2*sx*np.pi),'k-')
plt.axhline(0)
```

In this plot, we have shown the relationship with a line but in practice you will need to detect this because you will not know the true relationship in real data.

It is often hard to judge residual plots without prior experience, so it is helpful to repeatedly generate the plots above (without using a fixed seed so they are different every time. You can vary the generating mechanism by changing the sample size and fitted value distribution.

This artificial generation of plots is a good way to "calibrate" diagnostic plots. It is often hard to judge whether an apparent feature is real or just random variation. Repeated generation of plots under a known model, where there is or is not a violation of the assumption that the diagnostic plot is designed to check, is helpful in making this judgment.

It is also worthwhile to plot $\hat{\varepsilon}$ against x_i for potential predictors that are in the current model as well as those not used. The same considerations in these plots

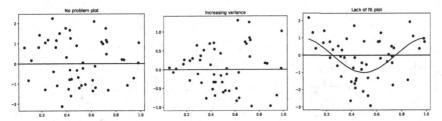

Figure 6.1   Residuals vs. fitted plots — the first suggests no change to the current model while the second shows nonconstant variance and the third indicates some nonlinearity, which should prompt some change in the structural form of the model.

should apply. For plots of residuals against predictors not in the model, any observed structure may indicate that this predictor should be included in the model.

We illustrate this using the savings dataset:

```
import faraway.datasets.savings
savings = faraway.datasets.savings.load()
lmod = smf.ols('sr ~ pop15 + pop75 + dpi + ddpi', savings).fit()
```

The basic plot, as seen in the first panel of Figure 6.2, shows residuals against fitted values. It is worth adding the $\hat{\varepsilon} = 0$ line to help with the interpretation.

```
plt.scatter(lmod.fittedvalues, lmod.resid)
plt.ylabel("Residuals")
plt.xlabel("Fitted values")
plt.axhline(0)
```

We see no cause for alarm in this plot. If we would like to examine the constant variance assumption more closely, it helps to plot $\sqrt{|\hat{\varepsilon}|}$ against $\hat{y}$. Once we have excluded the possibility of non-linearity in the first plot and the residuals look roughly symmetric, then we can effectively double the resolution by considering the absolute value of the residuals. For truly normal errors, $|\hat{\varepsilon}|$ would follow what is known as a half-normal distribution (since it has a density which is simply the upper half of a normal density). Such a distribution is quite skewed, but this effect can be reduced by the square root transformation.

```
plt.scatter(lmod.fittedvalues, np.sqrt(abs(lmod.resid)))
plt.ylabel("sqrt |Residuals|")
plt.xlabel("Fitted values")
```

The plot, as seen in the second panel of Figure 6.2, shows approximately constant variation. A quick numerical test to check nonconstant variance can be achieved using this regression where we check whether the size of $|\hat{\varepsilon}|$ is changing with the fitted values:

```
ddf = pd.DataFrame({'x':lmod.fittedvalues,
    'y':np.sqrt(abs(lmod.resid))})
dmod = smf.ols('y ~ x',ddf).fit()
dmod.sumary()
```

	coefs	stderr	tvalues	pvalues
Intercept	2.162	0.348	6.22	0.0000
x	-0.061	0.035	-1.77	0.0838

n=50 p=2 Residual SD=0.634 R-squared=0.06

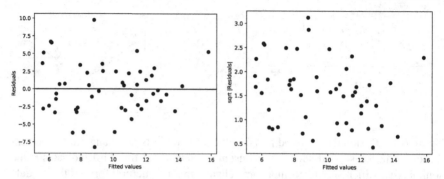

Figure 6.2   Residual vs. fitted plot for the savings data is shown on the left and the corresponding $\sqrt{|\hat{\epsilon}|}$ against $\hat{y}$ on the right. Both plots are satisfactory.

This test is not quite right, as some weighting should be used and the degrees of freedom should be adjusted, but there does not seem to be a clear problem with non-constant variance because the slope is not statistically significant. Another difficulty is that it only tests for a linear trend in the variation. A formal test may be good at detecting a particular kind of nonconstant variance, but may have no power to detect another. Residual plots are more versatile because unanticipated problems may be spotted.

A formal diagnostic test may have a reassuring aura of exactitude about it, but one needs to understand that any such test may be powerless to detect problems of an unsuspected nature. Graphical techniques are usually more effective at revealing structure that you may not have suspected. Of course, sometimes the interpretation of the plot may be ambiguous, but at least we can be sure that nothing is seriously wrong with the assumptions. For this reason, we usually prefer a graphical approach to diagnostics, with formal tests reserved for the clarification of indications discovered in the plots.

Now look at some residuals against predictor plots for the savings data:
```
plt.scatter(savings.pop15, lmod.resid)
plt.xlabel("%pop under 15")
plt.ylabel("Residuals")
plt.axhline(0)
plt.scatter(savings.pop75, lmod.resid)
plt.xlabel("%pop over 75")
plt.ylabel("Residuals")
plt.axhline(0)
```
The plots may be seen in Figure 6.3. Two groups can be seen in the first plot. Let's compare and test the variances in these groups. Given two independent samples from normal distributions, we can test for equal variance using the test statistic of the ratio of the two variances. The null distribution is an F with degrees of freedom given by the two samples:
```
numres = lmod.resid[savings.pop15 > 35]
denres = lmod.resid[savings.pop15 < 35]
fstat = np.var(numres,ddof=1)/np.var(denres,ddof=1)
```

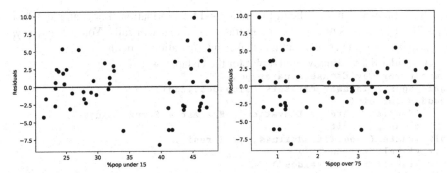

Figure 6.3   Residuals vs. predictor plots for the savings data.

```
2*(1-sp.stats.f.cdf(fstat,len(numres)-1,len(denres)-1))
 0.0136
```
A significant difference is seen.

When problems are seen in these diagnostic plots, some modification of the model is suggested. If some nonlinearity is observed, perhaps in conjunction with nonconstant variance, a transformation of the variables should be considered. The shape of the plot may suggest the transformation to be used. See Chapter 9 for more.

If the problem is solely one of nonconstant variance with no suggestion of non-linearity then the use of weighted least squares (see Section 8.1) may be appropriate. Alternatively, when nonconstant variance is seen in the plot of $\hat{\varepsilon}$ against $\hat{y}$, a transformation of the response y to $h(y)$ where $h()$ can be chosen so that var $h(y)$ is constant. To see how to choose h, consider this expansion:

$$
\begin{aligned}
h(y) &= h(Ey) + (y - Ey)h'(Ey) + \cdots \\
\text{var } h(y) &= 0 + h'(Ey)^2 \text{var } y + \cdots
\end{aligned}
$$

We ignore the higher order terms. For var $h(y)$ to be constant we need:

$$
h'(Ey) \propto (\text{var } y)^{-1/2}
$$

which suggests:

$$
h(y) = \int \frac{dy}{\sqrt{\text{var } y}} = \int \frac{dy}{\text{SD} y}
$$

For example if var $y = $ var $\varepsilon \propto (Ey)^2$, then $h(y) = \log y$ is suggested, while if var $\varepsilon \propto (Ey)$, then $h(y) = \sqrt{y}$.

In practice, you need to look at the plot of the residuals and fitted values and take a guess at the relationship. When looking at the plot, we see the change in $SD(y)$ rather than var y, because the SD is in the units of the response. If your initial guess is wrong, you will find that the diagnostic plot in the transformed scale is unsatisfactory. You can simply try another transformation — some experimentation is sensible.

Sometimes it can be difficult to find a good transformation. For example, when $y_i \leq 0$ for some i, square root or log transformations will fail. You can try, say, $\log(y + \delta)$, for some small δ but this makes interpretation difficult.

Consider the residual vs. fitted plot for the Galápagos data:

```
import faraway.datasets.galapagos
galapagos = faraway.datasets.galapagos.load()
gmod = smf.ols(
    'Species ~ Area + Elevation + Nearest + Scruz + Adjacent',
    galapagos).fit()
plt.scatter(gmod.fittedvalues, gmod.resid)
plt.ylabel("Residuals")
plt.xlabel("Fitted values")
plt.axhline(0)
```

We can see nonconstant variance (and evidence of nonlinearity) in the first plot of Figure 6.4. The square root transformation is often appropriate for count response

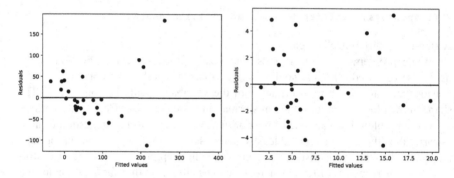

Figure 6.4   Residual vs. fitted plots for the Galápagos data before (left) and after (right) a square root transformation of the response.

data. The Poisson distribution is a good model for counts and that distribution has the property that the mean is equal to the variance, thus suggesting the square root transformation. We try that:

```
gmod = smf.ols(
    'np.sqrt(Species) ~ Area+Elevation+Nearest+Scruz+Adjacent',
    galapagos).fit()
plt.scatter(gmod.fittedvalues, gmod.resid)
plt.ylabel("Residuals")
plt.xlabel("Fitted values")
plt.axhline(0)
```

We see in the second plot of Figure 6.4 that the variance is now constant and the signs of nonlinearity have gone. Our guess at a variance stabilizing transformation worked out here, but had it not, we could always have tried something else.

6.1.2   Normality

The tests and confidence intervals we use are based on the assumption of normal errors. The residuals can be assessed for normality using a Q–Q plot. This compares the residuals to "ideal" normal observations. We plot the sorted residuals against $\Phi^{-1}(\frac{i}{n+1})$ for $i = 1, \ldots, n$. Let's try it out on the savings data:

```
sm.qqplot(lmod.resid, line="q")
```

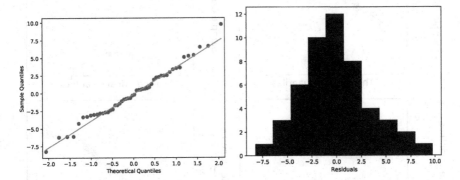

Figure 6.5   Normality checks for the savings data. A Q–Q plot is shown on the left and a histrogram on the right.

See the first plot of Figure 6.5 — qqline() adds a line joining the first and third quartiles. It is not influenced by outliers. Normal residuals should follow the line approximately. Here, the residuals look normal.

Histograms and boxplots are not suitable for checking normality:

```
plt.hist(lmod.resid)
plt.xlabel("Residuals")
```

The histogram seen in the second plot of Figure 6.5 does not have the expected bell shape. This is because we must group the data into bins. The choice of width and placement of these bins is problematic and the plot here is inconclusive.

We can get an idea of the variation to be expected in Q–Q plots in the following simulation. We generate data from different distributions:

1. Normal

2. Lognormal — an example of a skewed distribution

3. Cauchy — an example of a long-tailed (leptokurtic) distribution

4. Uniform — an example of a short-tailed (platykurtic) distribution

We generate each of these test cases as seen in Figure 6.6:

```
fig, axs = plt.subplots(2, 2, sharex=True)
sm.qqplot(np.random.normal(size=50),line="q",ax = axs[0,0])
sm.qqplot(np.exp(np.random.normal(size=50)),line="q",ax=axs[0,1])
sm.qqplot(np.random.standard_t(1,size=50),line="q",ax=axs[1,0])
sm.qqplot(np.random.sample(size=50),line="q",ax=axs[1,1])
plt.tight_layout()
```

It is not always easy to diagnose the problem in Q–Q plots. Sometimes extreme cases may be a sign of a long-tailed error like the Cauchy distribution or they can be just outliers. If removing such observations just results in other points becoming more prominent in the plot, the problem is likely due to a long-tailed error.

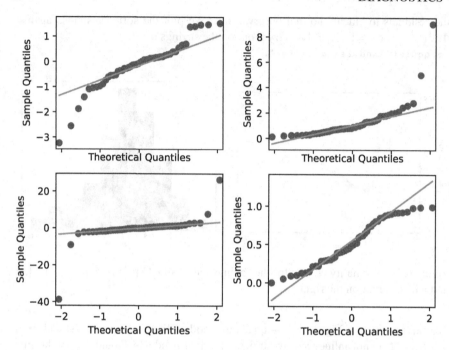

Figure 6.6   Q–Q plots of simulated data for, respectively, a skewed distribution, a long-tailed distribution and a short-tailed distribution.

When the errors are not normal, least squares estimates may not be optimal. They will still be best linear unbiased estimates, but other *robust* estimators may be more effective. Also tests and confidence intervals are not exact. However, we can appeal to the central limit theorem which will ensure that the tests and confidence intervals constructed will be increasingly accurate approximations for larger sample sizes. Hence, we can afford to ignore the issue, provided the sample is sufficiently large or the violation not particularly severe.

When nonnormality is found, the resolution depends on the type of problem found. For short-tailed distributions, the consequences of nonnormality are not serious and can reasonably be ignored. For skewed errors, a transformation of the response may solve the problem. For long-tailed errors, we might just accept the nonnormality and base the inference on the assumption of another distribution or use resampling methods such as the bootstrap or permutation tests. Alternatively, use robust methods, which give less weight to outlying observations but may again require resampling for the inference.

Also you may find that other diagnostics suggest changes to the model. In this changed model, the problem of nonnormal errors might not occur, so it is best to address any nonlinearity and nonconstant variance problems first.

The Shapiro–Wilk test is a formal test for normality:
```
sp.stats.shapiro(lmod.resid)
(0.987, 0.852)
```

The null hypothesis is that the residuals are normal. The first value is the statistic and the second the p-value. Since the p-value is large, we do not reject this hypothesis.

We can only recommend this in conjunction with a Q–Q plot at best. The p-value is not very helpful as an indicator of what action to take. With a large dataset, even mild deviations from nonnormality may be detected, but there would be little reason to abandon least squares because the effects of nonnormality are mitigated by large sample sizes. For smaller sample sizes, formal tests lack power.

6.1.3   Correlated Errors

It is difficult to check for correlated errors in general because there are just too many possible patterns of correlation that may occur. We do not have enough information to make any reasonable check. But some types of data have a structure which suggests where to look for problems. Data collected over time may have some correlation in successive errors. Spatial data may have correlation in the errors of nearby measurements. Data collected in blocks may show correlated errors within those blocks.

We illustrate the methods with an example of some temporal data. In general, the methods of time series analysis can be used but we will use methods more specific to the regression diagnostics problem. A combination of graphical and numerical methods is available.

The issue of global warming has attracted significant interest in recent years. Reliable records of annual temperatures taken with thermometers are only available back to the 1850s. Information about temperatures prior to this can be extracted from proxies such as tree rings. We can build a linear model to predict temperature since 1856 and then subsequently use this to predict earlier temperatures based on proxy information. The data we use here has been derived from Jones and Mann (2004). I have selected only proxies with mostly complete data. I have imputed some missing values and used some smoothing for ease of demonstration. Researchers seriously interested in historial climate should refer to the original data — see the R help page for details.

We start by fitting a model with the temperature as the response and the eight proxies as predictors. A full analysis would involve other steps, but we focus on the issue of correlated errors here.

```
import faraway.datasets.globwarm
globwarm = faraway.datasets.globwarm.load()
lmod=smf.ols('nhtemp ~ wusa + jasper + westgreen + chesapeake + \
    tornetrask + urals + mongolia + tasman', globwarm).fit()
```

There are missing values for nhtemp prior to 1856. The default behavior in R when performing a regression with missing values is to omit any case that contains a missing value. Hence this model uses only data from 1856 through 2000.

For temporal data such as these, it is sensible to plot the residuals against the time index. The plot is shown in the first panel of Figure 6.7. We have specified a missing valueless version of the data frame so that the years match up to the residuals.

```
plt.scatter(globwarm.year[lmod.resid.keys()], lmod.resid)
plt.axhline(0)
```

If the errors were uncorrelated, we would expect a random scatter of points above and below the $\varepsilon = 0$ line. In this plot, we see long sequences of points above or below the line. This is an indication of positive serial correlation. An alternative approach

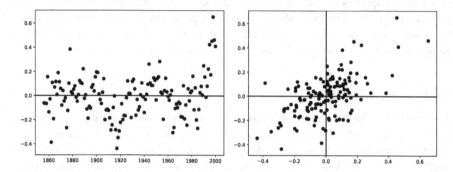

Figure 6.7   Diagnostic plots for correlated errors in the air quality data.

to checking for serial correlation is to plot successive pairs of residuals which we see in the second panel of Figure 6.7.
```
plt.scatter(lmod.resid.iloc[:-1],lmod.resid.iloc[1:])
plt.axhline(0,alpha=0.5)
plt.axvline(0,alpha=0.5)
```
We can see a positive correlation again indicating positive serial correlation. If you have some doubt as to the significance of the correlation, we can check with a test.

The Durbin–Watson test uses the statistic:

$$DW = \frac{\sum_{i=2}^{n}(\hat{\varepsilon}_i - \hat{\varepsilon}_{i-1})^2}{\sum_{i=1}^{n}\hat{\varepsilon}_i^2}$$

The null hypothesis is that the errors are uncorrelated. The null distribution based on the assumption of uncorrelated errors follows a linear combination of χ^2 distributions. The test is implemented in the lmtest package. The run test is an alternative.

We can compute the Durbin–Watson statistic:
```
sm.stats.stattools.durbin_watson(lmod.resid)
0.817
```
This is the statistic. A value of 2 is expected under the null, while less than 1 is a sign of a problem. In this example, we are not surprised to see that the errors are correlated. We do not expect that the proxies will model temperature perfectly and that higher or lower temperatures in one year might carry over to the next.

Sometimes, serial correlation can be caused by a missing covariate. For example, suppose that there is a quadratic relation between a predictor and the response but we use only a linear term in that predictor. The diagnostics will show serial correlation in the residuals but the real source of the problem is the missing quadratic term.

Although it is possible that difficulties with correlated errors can be removed by changing the structural part of the model, sometimes we must build the correlation directly into the model. This can be achieved using the method of generalized least squares — see Chapter 8.

6.2 Finding Unusual Observations

We may find that some observations do not fit the model well. Such points are called *outliers*. Other observations change the fit of the model in a substantive manner. These are called *influential* observations. It is possible for a point to have both these characteristics. A *leverage* point is extreme in the predictor space. It has the potential to influence the fit, but does not necessarily do so. It is important to first identify such points. Deciding what to do about them can be difficult.

6.2.1 Leverage

$h_i = H_{ii}$ are called *leverages* and are useful diagnostics. Since var $\hat{\varepsilon}_i = \sigma^2(1 - h_i)$, a large leverage, h_i, will make var $\hat{\varepsilon}_i$ small. The fit will be attracted toward y_i. Large values of h_i are due to extreme values in the X-space. h_i corresponds to a (squared) Mahalanobis distance defined by X which is $(x - \bar{x})^T \hat{\Sigma}^{-1}(x - \bar{x})$ where $\hat{\Sigma}$ is the estimated covariance of X. The value of h_i depends only on X and not y, so leverages contain only partial information about a case.

Since $\sum_i h_i = p$, an average value for h_i is p/n. A rough rule is that leverages of more than $2p/n$ should be looked at more closely.

We will use the savings dataset as an example here:

```
lmod = smf.ols('sr ~ pop15 + pop75 + dpi + ddpi', savings).fit()
diagv = lmod.get_influence()
hatv = pd.Series(diagv.hat_matrix_diag, savings.index)
hatv.sort_values().tail()
South Rhodesia     0.160809
Ireland            0.212236
Japan              0.223310
United States      0.333688
Libya              0.531457
```

We have shown the largest five leverages.

```
np.sum(hatv)
5.0
```

We verify that the sum of the leverages is indeed five — the number of parameters in the model.

Without making assumptions about the distributions of the predictors that would often be unreasonable, we cannot say how the leverages would be distributed. Nevertheless, we would like to identify unusually large values of the leverage. The half-normal plot is a good way to do this.

Half-normal plots are designed for the assessment of positive data. They could be used for $|\hat{\varepsilon}|$, but are more typically useful for diagnostic quantities like the leverages. The idea is to plot the data against the positive normal quantiles.

The steps are:

1. Sort the data: $x_{[1]} \leq \ldots x_{[n]}$.

2. Compute $u_i = \Phi^{-1}\left(\frac{n+i}{2n+1}\right)$.

3. Plot $x_{[i]}$ against u_i.

We do not usually expect a straight line relationship since we do not necessarily

expect a positive normal distribution for quantities like leverages. We are looking for
outliers, which will be apparent as points that diverge substantially from the rest of
the data.

We demonstrate the half-normal plot on the leverages for the savings data:
```
n=50
ix = np.arange(1,n+1)
halfq = sp.stats.norm.ppf((n+ix)/(2*n+1)),
plt.scatter(halfq, np.sort(hatv))
plt.annotate("Libya",(2.1,0.53))
plt.annotate("USA", (1.9,0.33))
```
The plot is the first shown in Figure 6.8. I have marked the countries with two
largest leverages. Leverages can also be used in scaling residuals. We have var $\hat{\varepsilon}_i =$

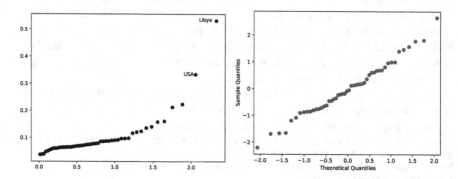

Figure 6.8   Half-normal plots for the leverages and a Q–Q plot for the standardized
residuals.

$\sigma^2(1 - h_i)$ which suggests the use of:

$$r_i = \frac{\hat{\varepsilon}_i}{\hat{\sigma}\sqrt{1 - h_i}}$$

You will find some variation in the names used by various authors and statistics
packages for this and other types of residual. This particular residual is sometimes
called the *standardized residual* or *Pearson residual*. If the model assumptions are
correct, var $r_i = 1$ and $corr(r_i, r_j)$ tend to be small. In statsmodels, the residual
found in lmod.resid_pearson only divides by $\hat{\sigma}$ and does not include the leverage
component. The more common definition of the Pearson residual can be obtained
as one of the objects returned from the get_influence() function and goes by the
name of *internally studentized residual* as seen below.

Standardized residuals are sometimes preferred in residual plots, as they have
been standardized to have equal variance. Standardization can only correct for the
natural nonconstant variance in residuals when the errors have constant variance.
If there is some underlying heteroscedasticity in the errors, standardization cannot
correct for it. We now compute and plot the standardized residuals for the savings
data:

```
rstandard = diagv.resid_studentized_internal
sm.qqplot(rstandard)
```

We have displayed the Q–Q plot of the standardized residuals in the second plot of Figure 6.8. Because these residuals have been standardized, we expect the points to approximately follow the $y = x$ line if normality holds. Another advantage of the standardized form is that we can judge the size easily. An absolute value of 2 would be large but not exceptional for a standardized residual, whereas a value of 4 would be very unusual under the standard normal.

Some authors recommend using standardized rather than raw residuals in all diagnostic plots. However, in many cases, the standardized residuals are not very different from the raw residuals except for the change in scale. Only when there are unusually large leverages will the differences be noticeable in the shape of the plot.

6.2.2   Outliers

An outlier is a point that does not fit the current model well. We need to be aware of such exceptions. An outlier test is useful because it enables us to distinguish between truly unusual observations and residuals that are large, but not exceptional. Outliers may or may not affect the fit substantially. We simulate some data to illustrate the possibilities:

```
np.random.seed(123)
testdata = pd.DataFrame({'x' : np.arange(1,11),
    'y' : np.arange(1,11) + np.random.normal(size=10)})
p1 = pd.DataFrame({'x': [5.5], 'y':[12]})
alldata = testdata.append(p1,ignore_index=True)
```

The first example adds an outlier with a central predictor value:

```
marksize = np.ones(11)
marksize[10] = 3
plt.scatter(alldata.x, alldata.y, s= marksize*5)
slope, intercept = np.polyfit(testdata.x, testdata.y,1)
plt.plot(testdata.x, intercept + slope * testdata.x)
slope, intercept = np.polyfit(alldata.x, alldata.y,1)
plt.plot(alldata.x, intercept + slope * alldata.x, '--')
```

The plot, seen in the first panel of Figure 6.9, shows a solid regression line for the fit without the additional point marked with a cross. The dashed line shows the fit with the extra point. There is not much difference. In particular, the slopes are very similar. This is an example of a point which is an outlier, but does not have large leverage or influence.

The second example introduces an extra point well outside the range of X:

```
p1 = pd.DataFrame({'x': [15], 'y':[15.1]})
alldata = testdata.append(p1,ignore_index=True)
plt.scatter(alldata.x, alldata.y, s= marksize*5)
slope, intercept = np.polyfit(testdata.x, testdata.y,1)
plt.plot(testdata.x, intercept + slope * testdata.x)
slope, intercept = np.polyfit(alldata.x, alldata.y,1)
plt.plot(alldata.x, intercept + slope * alldata.x, '--')
```

The plot, seen in the second panel of Figure 6.9, shows that the additional point makes little difference to the fitted regression line. This point has large leverage but is not an outlier and is not influential.

The third example puts the point in a different position on the response scale.

```
p1 = pd.DataFrame({'x': [15], 'y':[5.1]})
alldata = testdata.append(p1,ignore_index=True)
plt.scatter(alldata.x, alldata.y, s= marksize*5)
slope, intercept = np.polyfit(testdata.x, testdata.y,1)
plt.plot(testdata.x, intercept + slope * testdata.x)
slope, intercept = np.polyfit(alldata.x, alldata.y,1)
plt.plot(alldata.x, intercept + slope * alldata.x, '--')
```

The plot, seen in the third panel of Figure 6.9, shows that this additional point changes the regression line substantially. Although this point still has a large residual, the residuals for the other points are also made much larger with the addition of this point. Hence this point is both an outlier and an influential point. We must take care to discover such points because they can have a substantial effect on the conclusions of the analysis. Just looking at $\hat{\varepsilon}_i$ misses difficult observations, like that seen in

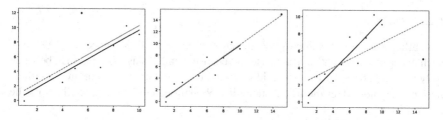

Figure 6.9   In all three plots, the additional point is marked with a cross. The solid line is the fit using the 10 original points and the dashed line is the fit with the added point.

the third example, which pull the regression line closer so that they conceal their true status. To detect such points, we exclude point i and recompute the estimates to get $\hat{\beta}_{(i)}$ and $\hat{\sigma}^2_{(i)}$ where (i) denotes that the ith case has been excluded. Hence:

$$\hat{y}_{(i)} = x_i^T \hat{\beta}_{(i)}$$

If $\hat{y}_{(i)} - y_i$ is large, then case i is an outlier. To judge the size of a potential outlier, we need an appropriate scaling. We find:

$$\text{vâr}\,(y_i - \hat{y}_{(i)}) = \hat{\sigma}^2_{(i)}(1 + x_i^T (X_{(i)}^T X_{(i)})^{-1} x_i)$$

and so we define the studentized (sometimes called jackknife or crossvalidated) residuals as:

$$t_i = \frac{y_i - \hat{y}_{(i)}}{\hat{\sigma}_{(i)}(1 + x_i^T (X_{(i)}^T X_{(i)})^{-1} x_i)^{1/2}}$$

which are distributed t_{n-p-1} if the model is correct and $\varepsilon \sim N(0, \sigma^2 I)$. Fortunately, there is an easier way to compute t_i:

$$t_i = \frac{\hat{\varepsilon}_i}{\hat{\sigma}_{(i)}\sqrt{1 - h_i}} = r_i \left(\frac{n-p-1}{n-p-r_i^2}\right)^{1/2}$$

which avoids doing n regressions.

In `statsmodels`, this residual is called the *externally studentized residual* and can be obtained as one of the objects returned from the `get_influence()` function as seen below.

Since $t_i \sim t_{n-p-1}$, we can calculate a p-value to test whether case i is an outlier. This is fine if we only test one preselected case. However, if we had $n = 100$ and tested all the cases, we would expect to find around five outliers using this procedure if we used a 5% significance level. Even though we might explicity test only one or two large t_is, we are implicitly testing all cases since we need to consider all the residuals to find out which ones are large. Some adjustment of the level of the test is necessary to avoid identifying an excess of outliers.

Suppose we want a level α test. Now P(all tests accept) = $1-$ P(at least one rejects) $\geq 1 - \sum_i P(\text{test } i \text{ rejects}) = 1 - n\alpha$. So this suggests that if an overall level α test is required, then a level α/n should be used in each of the tests. This method is called the *Bonferroni correction* and is used in contexts other than outliers. Its biggest drawback is that it is conservative — it finds fewer outliers than the nominal level of confidence would dictate. The larger that n is, the more conservative it gets.

Let's compute studentized residuals for the savings data and pick out the largest:

```
stud = pd.Series(diagv.resid_studentized_external, savings.index)
(pd.Series.idxmax(abs(stud)), np.max(abs(stud)))
('Zambia', 2.85)
```

The largest residual of 2.85 is rather large for a standard normal scale, but is it an outlier? Compute the Bonferroni critical value:

```
abs(sp.stats.t.ppf(0.05/(2*50),44))
3.53
```

Since 2.85 is less than 3.53, we conclude that Zambia is *not* an outlier. For simple regression, the minimum critical value occurs at $n = 23$ taking the value 3.51. This indicates that it is not worth the trouble of computing the outlier test p-value unless the studentized residual exceeds about 3.5 in absolute value.

Some points to consider about outliers:

1. Two or more outliers next to each other can hide each other.

2. An outlier in one model may not be an outlier in another when the variables have been changed or transformed. You will usually need to reinvestigate the question of outliers when you change the model.

3. The error distribution may not be normal and so larger residuals may be expected. For example, day-to-day changes in stock indices seem mostly normal, but larger changes occur from time to time.

4. Individual outliers are usually much less of a problem in larger datasets. A single point will not have the leverage to affect the fit very much. It is still worth identifying outliers if these types of observations are worth knowing about in the particular application. For large datasets, we need only to worry about clusters of outliers. Such clusters are less likely to occur by chance and more likely to represent actual structure. Finding these clusters is not always easy.

What should be done about outliers?

1. Check for a data-entry error first. These are relatively common. Unfortunately, the original source of the data may have been lost or may be inaccessible. If you

can be sure that the point is truly a mistake and was not actually observed, then the solution is simple: discard it.

2. Examine the physical context — why did it happen? Sometimes, the discovery of an outlier may be of singular interest. Some scientific discoveries spring from noticing unexpected aberrations. Another example of the importance of outliers is in the statistical analysis of credit card transactions. Outliers may represent fraudulent use. In such examples, the discovery of outliers is the main purpose of the modeling.

3. Exclude the point from the analysis, but try reincluding it later if the model is changed. The exclusion of one or more observations may make the difference between getting a statistically significant result or having some unpublishable research. This can lead to a difficult decision about what exclusions are reasonable. To avoid any suggestion of dishonesty, always report the existence of outliers even if you do not include them in your final model. Be aware that casual or dishonest exclusion of outliers is regarded as serious research malpractice.

4. Suppose you find outliers that cannot reasonably be identified as mistakes or aberrations, but are viewed as naturally occurring. Particularly if there are substantial numbers of such points, it is more efficient and reliable to use robust regression, as explained in Section 8.4. Routine outlier rejection in conjunction with least squares is not a good method of estimation. Some adjustment to the inferential methods is necessary in such circumstances. In particular, the uncertainty assessment for prediction needs to reflect the fact that extreme values can occur.

5. It is dangerous to exclude outliers in an automatic manner. National Aeronautics and Space Administration (NASA) launched the *Nimbus 7* satellite to record atmospheric information. After several years of operation in 1985, the British Antarctic Survey observed a large decrease in atmospheric ozone over the Antarctic. On further examination of the NASA data, it was found that the data processing program automatically discarded observations that were extremely low and assumed to be mistakes. Thus the discovery of the Antarctic ozone hole was delayed several years. Perhaps, if this had been known earlier, the chlorofluorocarbon (CFC) phaseout would have been agreed upon earlier and the damage could have been limited. See Stolarski et al. (1986) for more.

Here is an example of a dataset with multiple outliers. Data are available on the log of the surface temperature and the log of the light intensity of 47 stars in the star cluster CYG OB1, which is in the direction of Cygnus. These data appear in Rousseeuw and Leroy (1987). Read in and plot the data:

```
import faraway.datasets.star
star = faraway.datasets.star.load()
plt.scatter(star.temp, star.light)
```

There appears to be a positive correlation between temperature and light intensity, but there are four stars that do not fit the pattern. We fit a linear regression and add the fitted line to the plot:

```
lmod = smf.ols('light ~ temp',star).fit()
xr = np.array([np.min(star.temp),np.max(star.temp)])
plt.plot(xr, lmod.params[0] + lmod.params[1]*xr)
```

```
plt.xlabel("log(Temperature)")
plt.ylabel("log(Light Intensity)")
```
The plot is seen in Figure 6.10 with the regression line in solid type. This line does

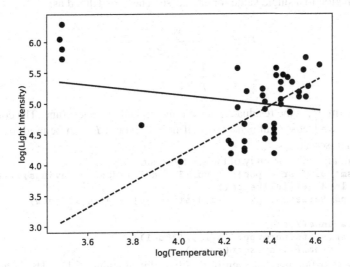

Figure 6.10   Regression line including four leftmost points is solid and excluding these points is dotted.

not follow the bulk of the data because it tries to fit the four unusual points. We check whether the outlier test detects these points:
```
stud = lmod.get_influence().resid_studentized_external
np.min(stud), np.max(stud)
(-2.049, 1.906)
```
No outliers are found even though we can see them clearly in the plot. The four stars on the upper left of the plot are giants. See what happens if these are excluded:
```
lmodr = smf.ols('light ~ temp',star[star.temp > 3.6]).fit()
plt.plot(xr, lmodr.params[0] + lmodr.params[1]*xr,'k--')
```
This illustrates the problem of multiple outliers. We can visualize the problems here and take corrective action, but for higher dimensional data this is much more difficult. Robust regression methods would be superior here. See Section 8.4.

6.2.3   Influential Observations

An influential point is one whose removal from the dataset would cause a large change in the fit. An influential point may or may not be an outlier and may or may not have large leverage, but it will tend to have at least one of these two properties. In Figure 6.9, the third panel shows an influential point but in the second panel, the added point is not influential.

There are several measures of influence. A subscripted (i) indicates the fit without case i. We might consider the change in the fit $X^T(\hat{\beta} - \hat{\beta}_{(i)}) = \hat{y} - \hat{y}_{(i)}$, but there

92   DIAGNOSTICS

will be *n* of these length *n* vectors to examine. For a more compact diagnostic, we might consider the change in the coefficients $\hat{\beta} - \hat{\beta}_{(i)}$. There will be $n \times p$ of these to look at. The Cook statistics are popular influence diagnostics because they reduce the information to a single value for each case. They are defined as:

$$D_i = \frac{(\hat{y} - \hat{y}_{(i)})^T (\hat{y} - \hat{y}_{(i)})}{p\hat{\sigma}^2}$$

$$= \frac{1}{p} r_i^2 \frac{h_i}{1 - h_i}$$

The first term, r_i^2, is the residual effect and the second is the leverage. The combination of the two leads to influence. A half-normal plot of D_i can be used to identify influential observations.

Continuing with our study of the savings data:
```
lmod = smf.ols('sr ~ pop15 + pop75 + dpi + ddpi', savings).fit()
diagv = lmod.get_influence()
cooks = pd.Series(diagv.cooks_distance[0], savings.index)
n=50
ix = np.arange(1,n+1)
halfq = sp.stats.norm.ppf((n+ix)/(2*n+1)),
plt.scatter(halfq, np.sort(cooks))
```
The Cook statistics may be seen in the first plot of Figure 6.11. The largest five values are:
```
cooks.sort_values().iloc[-5:]
Philippines    0.045221
Ireland        0.054396
Zambia         0.096633
Japan          0.142816
Libya          0.268070
```

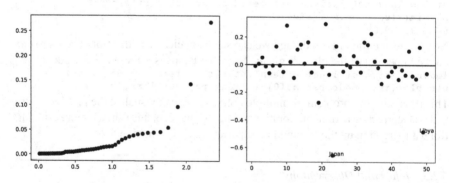

Figure 6.11   Half-normal plot of the Cook statistics and $\hat{\beta} - \hat{\beta}_{(i)}$'s for pop15 for the savings data.

We now exclude the largest one (Libya) and see how the fit changes:
```
lmodi = smf.ols('sr ~ pop15 + pop75 + dpi + ddpi',
    savings[cooks < 0.2]).fit()
pd.DataFrame({'with':lmod.params,'without':lmodi.params})
```

```
              with    without
Intercept  28.566087  24.524046
pop15      -0.461193  -0.391440
pop75      -1.691498  -1.280867
dpi        -0.000337  -0.000319
ddpi        0.409695   0.610279
```

Among other changes, we see that the coefficient for ddpi changed by about 50%. We do not like our estimates to be so sensitive to the presence of just one country. It would be rather tedious to leave out each country in turn, so we examine the leave-out-one differences in the coefficients:

```
p15d = diagv.dfbetas[:,1]
plt.scatter(np.arange(1,51),p15d)
plt.axhline(0)
ix = 22
plt.annotate(savings.index[ix],(ix, p15d[ix]))
ix = 48
plt.annotate(savings.index[ix],(ix, p15d[ix]))
```

We have plotted the change in the second parameter estimate, $(\hat{\beta}_{pop15})$ when a case is left out, as seen in the second panel of Figure 6.11. The identify function allows interactive identification of points by clicking the left mouse button on the plot and then using the middle mouse button to finish. This plot should be repeated for the other variables. Japan sticks out on this particular plot so we examine the effect of removing it:

```
lmodj = smf.ols('sr ~ pop15 + pop75 + dpi + ddpi',
      savings.drop(['Japan'])).fit()
lmodj.sumary()
```

```
           coefs  stderr  tvalues  pvalues
Intercept  23.940  7.784    3.08   0.0036
pop15      -0.368  0.154   -2.39   0.0210
pop75      -0.974  1.155   -0.84   0.4040
dpi        -0.000  0.001   -0.51   0.6112
ddpi        0.335  0.198    1.69   0.0987
```

```
n=49 p=5 Residual SD=3.738 R-squared=0.28
```

Comparing this to the full data fit, we observe several qualitative changes. Notice that the ddpi term is no longer significant and that the R^2 value has decreased a lot.

6.3   Checking the Structure of the Model

Diagnostics can also be used to detect deficiencies in the structural part of the model, given by $EY = X\beta$. The residuals are the best clues because signs of remaining systematic structure here indicate that something is wrong. A good diagnostic can often also suggest how the model can be improved. Formal lack of fit tests can be used in limited circumstances, usually requiring replication. See Section 8.3. But such tests, even if they are available, do not indicate how to improve the model.

We have already discussed plots of $\hat{\varepsilon}$ against $\hat{y}$ and x_i. We have used these plots to check the assumptions on the error, but they can also suggest transformations of the variables which might improve the structural form of the model.

We can also make plots of y against each x_i. Indeed these should be part of any exploratory analysis before model fitting begins. The drawback to these plots is that the other predictors often affect the relationship between a given predictor and the response. *Partial regression* or *added variable* plots can help isolate the effect of x_i on y. We regress y on all x except x_i, and get residuals $\hat{\delta}$. These represent y with the other X-effect taken out. Similarly, if we regress x_i on all x except x_i, and get residuals $\hat{\gamma}$, we have the effect of x_i with the other X-effect taken out. The added variable plot shows $\hat{\delta}$ against $\hat{\gamma}$. Look for nonlinearity and outliers and/or influential observations in the plot.

We illustrate using the savings dataset as an example again. We construct a partial regression (added variable) plot for pop15:

```
d = smf.ols('sr ~ pop75 + dpi + ddpi', savings).fit().resid
m = smf.ols('pop15 ~ pop75 + dpi + ddpi', savings).fit().resid
plt.scatter(m,d)
plt.xlabel("pop15 residuals")
plt.ylabel("sr residuals")
plt.plot([-10,8], [-10*lmod.params[1], 8*lmod.params[1]])
```

The plot, shown in the first panel of Figure 6.12, shows nothing remarkable. There is no sign of nonlinearity or unusual points. An interesting feature of such plots is revealed by considering the regression line. We show this on the plot and notice that it is the same as the corresponding coefficient from the full regression:

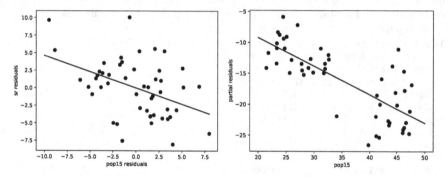

Figure 6.12   Partial regression (left) and partial residual (right) plots for the savings data.

```
np.polyfit(m,d,deg=1)[0], lmod.params[1]
-0.461, -0.461
```

The partial regression plot provides some intuition about the meanings of regression coefficients. We are looking at the marginal relationship between the response and the predictor after the effect of the other predictors has been removed. Multiple regression is difficult because we cannot visualize the full relationship because of the high dimensionality. The partial regression plot allows us to focus on the relationship between one predictor and the response, much as in simple regression.

Partial residual plots serve as an alternative to added variable plots. We construct the response with the predicted effect of the other X removed:

$$y - \sum_{j \neq i} x_j \hat{\beta}_j = \hat{y} + \hat{\varepsilon} - \sum_{j \neq i} x_j \hat{\beta}_j = x_i \hat{\beta}_i + \hat{\varepsilon}$$

The partial residual plot is then $\hat{\varepsilon} + \hat{\beta}_i x_i$ against x_i. Again the slope on the plot will be $\hat{\beta}_i$ and the interpretation is the same. Partial residual plots are believed to be better for nonlinearity detection while added variable plots are better for outlier/influential detection.

A partial residual plot can be constructed directly. (In some version of this plot, the partial residuals are centered but this just shifts the scale on the vertical axis).

```
pr = lmod.resid + savings.pop15*lmod.params[1]
plt.scatter(savings.pop15, pr)
plt.xlabel("pop15")
plt.ylabel("partial residuals")
plt.plot([20,50], [20*lmod.params[1], 50*lmod.params[1]])
```

We see two groups in the plot in the second panel of Figure 6.12. It suggests that there may be a different relationship in the two groups. We investigate this:

```
smf.ols('sr ~ pop15 + pop75 + dpi + ddpi',
        savings[savings.pop15 > 35]).fit().summary()
```

	coefs	stderr	tvalues	pvalues
Intercept	-2.434	21.155	-0.12	0.9097
pop15	0.274	0.439	0.62	0.5408
pop75	-3.548	3.033	-1.17	0.2573
dpi	0.000	0.005	0.08	0.9339
ddpi	0.395	0.290	1.36	0.1896

n=23 p=5 Residual SD=4.454 R-squared=0.16

```
smf.ols('sr ~ pop15 + pop75 + dpi + ddpi',
        savings[savings.pop15 < 35]).fit().summary()
```

	coefs	stderr	tvalues	pvalues
Intercept	23.962	8.084	2.96	0.0072
pop15	-0.386	0.195	-1.98	0.0609
pop75	-1.328	0.926	-1.43	0.1657
dpi	-0.000	0.001	-0.63	0.5326
ddpi	0.884	0.295	2.99	0.0067

n=27 p=5 Residual SD=2.772 R-squared=0.51

In the first regression on the subset of underdeveloped countries, we find no relation between the predictors and the response. We know from our previous examination of these data that this result is not attributable to outliers or unsuspected transformations. In contrast, there is a strong relationship in the developed countries. The strongest predictor is growth together with some relationship to proportion of population under 15. This latter effect has been reduced from prior analyses because we have reduced the range of this predictor by the subsetting operation. The graphical analysis has shown a relationship in the data that a purely numerical analysis might easily have missed.

Sometimes it can be helpful to introduce extra dimensions into diagnostic plots with the use of color, plotting symbol or size. Alternatively, plots can be faceted. The ggplot2 package is convenient for this purpose. Here are a couple of examples:

```
savings['age'] = np.where(savings.pop15 > 35, 'young', 'old')
sns.lmplot('ddpi','sr',data=savings, hue='age',legend_out=False)
```

or

```
sns.lmplot('ddpi','sr',data=savings, col='age')
```

The plots in Figure 6.13 show how we can distinguish the two levels of the status variable derived from the population proportion under 15 in two different ways. In

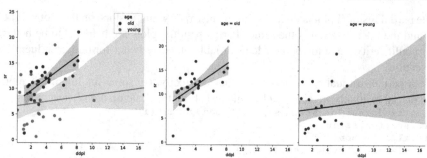

Figure 6.13   Introducing another dimension to diagnostic plots. Shape is used to denote the status variable on the left while faceting is used for the two right panels.

this case, the second set of plots is more effective. Note how we have added a regression line with 95% confidence bands, making it clear how the relationship differs between the two groups.

Higher dimensional plots can also be useful for detecting structures that cannot be seen in two dimensions. These are usually interactive in nature so you need to try them to see how they work. We can make three-dimensional plots where color, point size and rotation are used to give the illusion of a third dimension. We can also link two or more plots so that points which are selected in one plot are highlighted in another. These tools look good but it is not clear whether they actually are useful in practice. There are communication difficulties, as these plots cannot be easily printed.

Nongraphical techniques for checking the structural form of the model usually involve proposing alternative transformations or recombinations of the variables. This approach is explored in Chapter 9.

6.4   Discussion

Some assumptions are more important than others because some violations are more serious in that they can lead to very misleading conclusions. We can order these assumptions according to their importance:

1. The systematic form of the model. If you get this seriously wrong, then predictions will be inaccurate and any explanation of the relationship between the variables may be biased in misleading ways.

2. Dependence of errors. The presence of strong dependence means that there is less information in the data than the sample size may suggest. Furthermore, there is a risk that the analyst will mistakenly introduce systematic components to the model in an attempt to deal with an unsuspected dependence in the errors. Unfortunately, it is difficult to detect dependence in the errors using regression diagnostics except in special situations such as temporal data. For other types of data, the analyst will need to rely on less testable assumptions about independence based on contextual knowledge.

3. Nonconstant variance. A failure to address this violation of the linear model assumptions may result in inaccurate inferences. In particular, prediction uncertainty may not be properly quantified. Even so, excepting serious violations, the adequacy of the inference may not be seriously compromised.

4. Normality. This is the least important assumption. For large datasets, the inference will be quite robust to a lack of normality as the central limit theorem will mean that the approximations will tend to be adequate. Unless the sample size is quite small or the errors very strongly abnormal, this assumption is not crucial to success.

Although it is not part of regression diagnostics, it is worthwhile also mentioning that an even more important assumption is that the data at hand are relevant to the question of interest. This requires some qualitative judgment and is not checkable by plots or tests.

Excercises

1. Using the sat dataset, fit a model with the total SAT score as the response and expend, salary, ratio and takers as predictors. Perform regression diagnostics on this model to answer the following questions. Display any plots that are relevant. Do not provide any plots about which you have nothing to say. Suggest possible improvements or corrections to the model where appropriate.

   (a) Check the constant variance assumption for the errors.
   (b) Check the normality assumption.
   (c) Check for large leverage points.
   (d) Check for outliers.
   (e) Check for influential points.
   (f) Check the structure of the relationship between the predictors and the response.

2. Using the teengamb dataset, fit a model with gamble as the response and the other variables as predictors.

   (a) Produce the residual-fitted plot and comment.
   (b) Check the normality assumption. Is there an outlier?
   (c) Compute the two forms of the standardized residual. One where the raw residual is normalized only by the scale and one where the raw residual is normalized using both the scale and the leverage. Plot the ratio of these two residuals against the index of observation. Comment on the variation of this ratio.
   (d) Compute the standard deviation of the two types of residual discussed in the previous question. What value is expected?
   (e) Compute the (externally) studentized residual and make a QQ plot. Apart from the change in scale on the y-axis, is there any difference between this plot and the one from (b)?
   (f) Compute the Bonferroni cut-off and use it to identify the outliers in this model.

(g) Determine a transform of the response which results in a satisfactory residual-fitted plot.

3. For the `prostate` data, fit a model with `lpsa` as the response and the other variables as predictors.

(a) Make the residual-fitted plot and comment.

(b) Check for large leverage points.

(c) Display the regression summary and identify all the predictors which are not statistically significant. Fit a model with all these predictors removed. Construct the residual-fitted plot and compare to that for the full model.

(d) Recompute the leverages and plot. Identify any unusually large leverages.

(e) For any unusual cases identified in the previous question, display the predictor values converted to standard units. Which predictors made these cases unusual?

(f) Compare the regression summaries for the reduced model with and without any points identified as having large leverage.

4. For the `swiss` data, fit a model with `Fertility` as the response and the other variables as predictors. Note that you will need to change one of the variable names. In R, six diagnostic plots are readily available for a linear model. In this question, you will reproduce all six.

(a) Make a residual-fitted plot. Overlay a smooth fitted curve using lowess.

(b) Make a QQ plot of the standardized residuals.

(c) A 'scale-location' plot. This has the square root of the absolute standardized residuals on the vertical, the fitted values on the horizontal with an overlaid lowess smooth.

(d) On the horizontal axis, plot the index of the observations. At each observation, draw a vertical bar to the height of the Cook's statistic. Label the largest Cook's statistic.

(e) Plot the standardized residuals on the vertical and the leverages on the horizontal. Overlay a lowess smoothed line.

(f) Plot the Cook's distance on the vertical and h/(1-h) (where h is the leverage) on the horizontal.

5. Using the `cheddar` data, fit a model with `taste` as the response and the other three variables as predictors.

(a) Make a plot of acetic against taste showing the univariate regression line on the plot along with a confidence band. Does acetic appear significant in predicting taste?

(b) Make the partial residual plot for acetic with respect to the full model with all three predictors. Show a confidence band on the plot. Contrast the plot with that from the previous question. Under what circumstances would these plots present a similar configuration of points?

(c) Produce the partial residual plot for H2S. Interpret the plot.

(d) Compute the predicted value of taste for H2S varying from its minimum to its maximum value while the other two predictors are held fixed at their mean values. Plot the resulting prediction line on top of a scatterplot of H2S and taste. What is this kind of plot called? How does it compare with the partial residual plot of the previous question?

(e) Produce the partial regression plot for H2S. Comment and contrast with the previous two plots.

6. Using the mba data, fit a model with happy as the response and the other four variables as predictors.

(a) Construct the residual-fitted plot. Do the linear model assumptions appear to be broken? Why are there diagnoal lines of points on the plot?

(b) Construct the QQ-plot and comment.

(c) Two rows in the dataset are identical. Identify these two rows. Does this indicate that the linear model assumptions are broken?

(d) Construct the partial residual plot for sex. Does this plot have any practical value?

(e) Compute the mean of the partial residuals (from the previous question) when sex=1. Why is this value the same as the regression coefficient for sex?

7. Using the tvdoctor data, fit a model with life as the response and the other two variables as predictors.

(a) Make scatterplots with life as the response and each of the predictors. Show the univariate regression line (but no confidence band) on each plot. Comment.

(b) Construct the partial residual plots for both predictors. Contrast these with the two plots of the previous question.

(c) Use a log transform on both predictors and refit the model. Present both partial residual plots and comment.

(d) How helpful were partial residual plots in finding the best transformation on the predictors in this example?

8. For the divusa data, fit a model with divorce as the response and the other variables, except year as predictors. Check for serial correlation.

9. Consider a sequence of regression models with only one predictor where the sample size takes integer value between 10 and 50. Compute the Bonferroni critical value for each of these models. Make a plot of your findings. What was the minimum observed value of the critical value and for what sample size did it occur?

Chapter 7

Problems with the Predictors

7.1 Errors in the Predictors

The regression model $Y = X\beta + \varepsilon$ allows for Y being measured with error by having the ε term, but what if the X is measured with error? In other words, what if the X we see is not exactly the X that potentially drives Y? It is not unreasonable that there might be errors in measuring X. For example, consider the problem of determining the effects of being exposed to a potentially hazardous substance such as second-hand tobacco smoke. Such exposure would be a predictor in such a study, but it is very hard to measure this exactly over a period of years.

One should not confuse errors in predictors with treating X as a random variable. For observational data, X could be regarded as a random variable, but the regression inference proceeds conditionally on a fixed value for X. For example, suppose we conducted a study where the blood sugar level of the participants was a predictor. Since the participants are sampled from a larger population, the blood sugar level is a random variable. However, once we select the participants and measure the blood sugar levels, these become fixed for the purposes of the regression. But our interest in this section is the situation where we are not able to measure the blood sugar levels accurately. In this case, the value we record will not be the same as the value that possibly affects the response.

Suppose that what we observe is (x_i^O, y_i^O) for $i = 1, \ldots n$ which are related to the true values (x_i^A, y_i^A):

$$
\begin{aligned}
y_i^O &= y_i^A + \varepsilon_i \\
x_i^O &= x_i^A + \delta_i
\end{aligned}
$$

where the errors ε and δ are independent. The situation is depicted in Figure 7.1. The true underlying relationship is:

$$y_i^A = \beta_0 + \beta_1 x_i^A$$

but we only see (x_i^O, y_i^O). Putting it together, we get:

$$y_i^O = \beta_0 + \beta_1 x_i^O + (\varepsilon_i - \beta_1 \delta_i)$$

Suppose we use least squares to estimate β_0 and β_1. Let's assume $E\varepsilon_i = E\delta_i = 0$ and that var $\varepsilon_i = \sigma_\varepsilon^2$, var $\delta_i = \sigma_\delta^2$. Let:

$$\sigma_x^2 = \sum (x_i^A - \bar{x}^A)^2 / n \qquad \sigma_{x\delta} = cov(x^A, \delta)$$

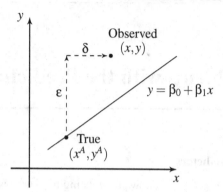

Figure 7.1   Measurement error: true vs. observed data.

For observational data, σ_x^2 is (almost) the sample variance of X^A while for a controlled experiment we can view it as just a numerical measure of the spread of the design. A similar distinction should be made for $cov(x^A, \delta)$ although in many cases, it will be reasonable to assume that this is zero.

Now $\hat{\beta}_1 = \sum(x_i - \bar{x})y_i / \sum(x_i - \bar{x})^2$ and after some calculation we find that:

$$E\hat{\beta}_1 = \beta_1 \frac{\sigma_x^2 + \sigma_{x\delta}}{\sigma_x^2 + \sigma_\delta^2 + 2\sigma_{x\delta}}$$

There are two main special cases of interest:

1. If there is no relation between X^A and δ, $\sigma_{x\delta} = 0$, this simplifies to:

$$E\hat{\beta}_1 = \beta_1 \frac{1}{1 + \sigma_\delta^2/\sigma_x^2}$$

So $\hat{\beta}_1$ will be biased toward zero, regardless of the sample size. If σ_δ^2 is small relative to σ_x^2, then the problem can be ignored. In other words, if the variability in the errors of observation of X is small relative to the range of X, then we need not be too concerned. For multiple predictors, the usual effect of measurement errors is also to bias the $\hat{\beta}$ in the direction of zero.

2. In controlled experiments, we need to distinguish two ways in which error in x may arise. In the first case, we measure x so although the true value is x^A, we observe x^O. If we were to repeat the measurement, we would have the same x^A, but a different x^O. In the second case, you fix x^O — for example, you make up a chemical solution with a specified concentration x^O. The true concentration would be x^A. Now if you were to repeat this, you would get the same x^O, but the x^A would be different. In this latter case we have:

$$\sigma_{x\delta} = cov(X^O - \delta, \delta) = -\sigma_\delta^2$$

and then we would have $E\hat{\beta}_1 = \beta_1$. So our estimate would be unbiased. This seems paradoxical, until you notice that the second case effectively reverses the roles of x^A and x^O and if you get to observe the true X, then you will get an unbiased estimate of β_1. See Berkson (1950) for a discussion of this.

If the model is used for prediction purposes, we can make the same argument as in the second case above. In repeated "experiments," the value of x at which the prediction is to be made will be fixed, even though these may represent different underlying "true" values of x.

In cases where the error in X can simply not be ignored, we should consider alternatives to the least squares estimation of β. The simple least squares regression equation can be written as:

$$\frac{y - \bar{y}}{SD_y} = r\frac{(x - \bar{x})}{SD_x}$$

so that $\hat{\beta}_1 = rSD_y/SD_x$. Note that if we reverse the roles of x and y, we do not get the same regression equation. Since we have errors in both x and y in our problem, we might argue that neither one, in particular, deserves the role of response or predictor and so the equation should be the same either way. One way to achieve this is to set $\hat{\beta}_1 = SD_y/SD_x$. This is known as the *geometric mean functional relationship*. More on this can be found in Draper and Smith (1998). Another approach is to use the SIMEX method of Cook and Stefanski (1994), which we illustrate below.

Before starting, we load in the packages for this chapter:

```
import pandas as pd
import numpy as np
import matplotlib.pyplot as plt
import scipy as sp
import statsmodels.api as sm
import statsmodels.formula.api as smf
import faraway.utils
```

Consider some data on the speed and stopping distances of cars in the 1920s. We plot the data, as seen in Figure 7.2, and fit a linear model:

```
import faraway.datasets.cars
cars = faraway.datasets.cars.load()
est = np.polyfit(cars.speed, cars.dist, 1)
est.round(2)
array([  3.93, -17.58])
```

The `np.polyfit()` function fits polynomials to data. In this case, we want only a linear term, so we specify the order as "1". This simple function returns only the coefficients but in a different order from the more full-featured `statsmodels` function. The coefficients are in descending order so that the linear term is first and the constant term second.

We could explore transformations and diagnostics for these data, but we will focus on just the measurement error issue. Now we investigate the effect of adding measurement error to the predictor. We start by plotting the original linear fit to the data and experiment with adding differing amounts of error to the predictors. We do this by varying the standard deviation of δ over $1, 2, 5$. We have set the random seed for repeatability. The fits are shown in Figure 7.2:

```
fig, ax = plt.subplots()
ax.scatter(cars.speed, cars.dist, label=None)
```

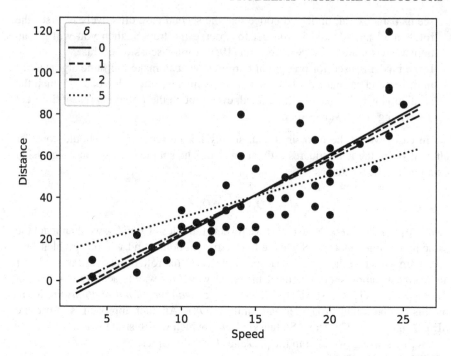

Figure 7.2 Stopping distances and speeds of cars. The least squares fit is shown as a solid line. The fits with three progressively larger amounts of measurement error on the speed are shown as dotted lines, where the slope gets shallower as the error increases.

```
plt.xlabel("Speed")
plt.ylabel("Distance")
xr = np.array(ax.get_xlim())
np.random.seed(123)
ax.plot(xr, est[1] + est[0] * xr,label="0")
est1 = np.polyfit(cars.speed + np.random.normal(size=50),
    cars.dist, 1)
ax.plot(xr, est1[1] + est1[0] * xr, 'k--',label="1")
est2 = np.polyfit(cars.speed + np.random.normal(scale=2,size=50),
    cars.dist, 1)
ax.plot(xr, est2[1] + est2[0] * xr, 'k-.',label="2")
est5 = np.polyfit(cars.speed + np.random.normal(scale=5,size=50),
    cars.dist, 1)
ax.plot(xr, est5[1] + est5[0] * xr, 'k:',label="5")
plt.legend(title='$\delta$')
```

We can see that the slope becomes shallower as the amount of noise increases. We extract the four sets of parameter estimates, est, est1, est2, est5 and print them as a neat table:

```
ee = pd.DataFrame.from_records([est, est1, est2, est5],
    columns=["slope","intercept"])
```

```
ee.insert(0,"SDdelta",[0,1,2,5])
print(ee.round(2).to_string(index=False))
SDdelta  slope  intercept
      0   3.93     -17.58
      1   3.76     -14.98
      2   3.47     -10.77
      5   2.06       9.95
```

We see how the slope decreases as the noise increases. A side effect of this is that the intercept increases.

Suppose we knew that the predictor, speed, in the original data had been measured with a known error variance, say 0.5. Given what we have seen in the simulated measurement error models, we might extrapolate back to suggest an estimate of the slope under no measurement error. This is the idea behind SIMEX.

Here we simulate the effects of adding normal random error with variances ranging from 0.1 to 0.5, replicating the experiment 1000 times for each setting:

```
vv = np.repeat(np.array([0.1,  0.2,  0.3,  0.4,  0.5]),
    [1000,  1000,  1000,  1000,  1000])
slopes = np.zeros(5000)
for i in range(5000):
    slopes[i] = np.polyfit(cars.speed+np.random.normal(
        scale=np.sqrt(vv[i]),size=50), cars.dist, 1)[0]
```

Now we plot the mean slopes for each variance as seen in Figure 7.3. We are assuming that the data have variance 0.5 so the extra variance is added to this:

```
betas = np.reshape(slopes, (5, 1000)).mean(axis=1)
betas = np.append(betas,est[0])
variances = np.array([0.6, 0.7, 0.8, 0.9, 1.0, 0.5])
gv = np.polyfit(variances, betas,1)
plt.scatter(variances, betas)
xr = np.array([0,1])
plt.plot(xr, np.array(gv[1] + gv[0]*xr))
plt.plot([0], [gv[1]], marker='x', markersize=6)
```

We examine the fit:

```
gv.round(2)
array([-0.13,  3.99])
```

The predicted value of $\hat{\beta}$ at variance equal to zero, that is, no measurement error, is 3.99. Better models for extrapolation are worth considering; see Cook and Stefanski (1994) for details.

7.2 Changes of Scale

Sometimes we want to change the scale of the variables. Perhaps we need to change the units of measurement, say from inches to centimeters. A change of scale is often helpful when variables take values which are all very large or all very small. For example, we might find $\hat{\beta} = 0.000000351$. It can be misleading to deal with quantities like this because it's easy to lose track of the number of leading zeroes. Humans deal better with interpreting moderate-sized numbers like $\hat{\beta} = 3.51$. So to avoid confusing yourself or your readers, a change of scale can be beneficial.

In more extreme cases, a change of scale can be needed to ensure numerical stability. Although many algorithms try to be robust against variables of widely

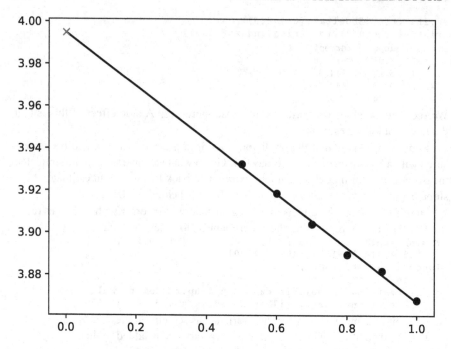

Figure 7.3   Simulation–extrapolation estimation of the unbiased slope in the presence of measurement error in the predictors. We predict $\hat{\beta} = 4.0$ at a variance of zero.

different scales, it is possible to run into calculation errors. Most methods work more reliably when variables are on roughly similar scales.

Suppose we reexpress x_i as $(x_i + a)/b$. Rescaling x_i leaves the t- and F-tests and $\hat{\sigma}^2$ and R^2 unchanged and $\hat{\beta}_i \to b\hat{\beta}_i$. Rescaling y in the same way leaves the t- and F-tests and R^2 unchanged, but $\hat{\sigma}$ and $\hat{\beta}$ will be rescaled by b.

To demonstrate this, we use the savings data:

```
import faraway.datasets.savings
savings = faraway.datasets.savings.load()
lmod = smf.ols('sr ~ pop15 + pop75 + dpi + ddpi', savings).fit()
lmod.sumary()
```

	coefs	stderr	tvalues	pvalues
Intercept	28.566	7.355	3.88	0.0003
pop15	-0.461	0.145	-3.19	0.0026
pop75	-1.691	1.084	-1.56	0.1255
dpi	-0.000	0.001	-0.36	0.7192
ddpi	0.410	0.196	2.09	0.0425

n=50 p=5 Residual SD=3.803 R-squared=0.34

The coefficient for income is rather small — let's measure income in thousands of dollars instead and refit:

```
lmod = smf.ols('sr ~ pop15 + pop75 + I(dpi/1000) + ddpi',
```

```
        savings).fit()
lmod.sumary()
              coefs stderr tvalues pvalues
Intercept     28.566  7.355    3.88  0.0003
pop15         -0.461  0.145   -3.19  0.0026
pop75         -1.691  1.084   -1.56  0.1255
I(dpi / 1000) -0.337  0.931   -0.36  0.7192
ddpi           0.410  0.196    2.09  0.0425
```

n=50 p=5 Residual SD=3.803 R-squared=0.34

What changed and what stayed the same?

One rather thorough approach to scaling is to convert all the variables to standard units (mean 0 and variance 1) using the scale() command:

```
scsav = savings.apply(sp.stats.zscore)
lmod = smf.ols('sr ~ pop15 + pop75 + dpi + ddpi', scsav).fit()
lmod.sumary()
          coefs stderr tvalues pvalues
Intercept 0.000  0.121    0.00  1.0000
pop15     -0.942  0.295   -3.19  0.0026
pop75     -0.487  0.312   -1.56  0.1255
dpi       -0.075  0.206   -0.36  0.7192
ddpi       0.262  0.126    2.09  0.0425
```

n=50 p=5 Residual SD=0.857 R-squared=0.34

As may be seen, the intercept is zero. This is because the regression plane always runs through the point of the averages, which because of the centering, is now at the origin. Such scaling has the advantage of putting all the predictors and the response on a comparable scale, which makes comparisons simpler. It also allows the coefficients to be viewed as a kind of partial correlation — the values will always be between minus one and one. It also avoids some numerical problems that can arise when variables are of very different scales. The interpretation effect of this scaling is that the regression coefficients now represent the effect of a one standard unit increase in the predictor on the response in standard units — this might or might not be easy to interpret.

When the predictors are on comparable scales, it can be helpful to construct a plot of the estimates with confidence intervals as seen in Figure 7.4.

```
edf = pd.concat([lmod.params, lmod.conf_int()],axis=1).iloc[1:,]
edf.columns = ['estimate','lb','ub']
npreds = edf.shape[0]
fig, ax = plt.subplots()
ax.scatter(edf.estimate,np.arange(npreds))
for i in range(npreds):
    ax.plot([edf.lb[i], edf.ub[i]], [i, i])
ax.set_yticks(np.arange(npreds))
ax.set_yticklabels(edf.index)
ax.axvline(0)
```

In the presence of binary predictors, scaling might be done differently. For example, we notice that the countries in the savings data divide into two clusters based on age. We can set a division at 35% for pop15:

```
savings['age'] = np.where(savings.pop15 > 35, 0, 1)
```

so that younger countries are coded as zero and older countries as one. A binary

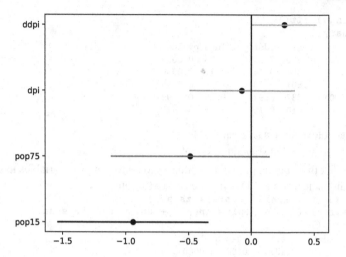

Figure 7.4   Plot of the standardized coefficients for the savings model; 95% confidence intervals are shown.

predictor taking the values of 0/1 with equal probability has a standard deviation of one half. This suggests scaling the other continuous predictors by two SDs rather than one:

```
savings['dpis'] = sp.stats.zscore(savings.dpi)/2
savings['ddpis'] = sp.stats.zscore(savings.ddpi)/2
smf.ols('sr ~ age + dpis + ddpis', savings).fit().sumary()
```

	coefs	stderr	tvalues	pvalues
Intercept	6.818	1.011	6.75	0.0000
age	5.284	1.585	3.33	0.0017
dpis	-1.549	1.593	-0.97	0.3361
ddpis	2.443	1.097	2.23	0.0309

```
n=50 p=4 Residual SD=3.800 R-squared=0.32
```

Now the interpretation of the coefficients becomes easier. The predicted difference between older and younger countries is a savings rate 5.28% higher in the former. This is a difference of two standard deviations. The same change of two standard deviations in ddpi means a difference of one in the new scale of ddpis. So we could say that a typical country with a high growth rate has a savings rate 2.47% higher than one with a low growth rate. Now ddpi is really a continuous variable so this interpretation is just for convenient intuition. Another way to achieve a similar effect is to use a $-1/+1$ coding rather than 0/1 so that the standard scaling can be used on the continuous predictors. See Gelman (2008) for more details.

7.3 Collinearity

When some predictors are linear combinations of others, then $X^T X$ is singular, and we have a lack of identifiability as discussed in Section 2.10. Another name for this problem is exact collinearity. There is no unique least squares estimate of β. The solution may require removing some predictors.

A more challenging problem arises when $X^T X$ is close to singular but not exactly so. This is known as collinearity or sometimes, multicollinearity. Collinearity leads to imprecise estimates of β. The signs of the coefficients can be the opposite of what intuition about the effect of the predictor might suggest. The standard errors are inflated so that t-tests may fail to reveal significant factors. The fit becomes very sensitive to measurement errors where small changes in y can lead to large changes in $\hat{\beta}$.

Collinearity can be detected in several ways:

1. Examination of the correlation matrix of the predictors may reveal values close to -1 or $+1$ indicating large *pairwise* collinearities.

2. A regression of x_i on all other predictors gives R_i^2. R_i^2 close to one indicates a problem because it means one predictor can almost be predicted exactly by a linear combination of other predictors. Repeat for all predictors. The offending linear combination may be discovered by examining the regression coefficients of these regressions.

3. Examine the eigenvalues of $X^T X$, $\lambda_1 \geq \cdots \geq \lambda_p \geq 0$. Zero eigenvalues denote exact collinearity, while the presence of some small eigenvalues indicates multi-collinearity. The condition number κ measures the relative sizes of the eigenvalues and is defined as:

$$\kappa = \sqrt{\frac{\lambda_1}{\lambda_p}}$$

where $\kappa \geq 30$ is considered large. Other condition numbers, $\sqrt{\lambda_1/\lambda_i}$ are also worth considering because they indicate whether more than just one independent linear combination is to blame. Alternative calculations involve standardizing the predictors and/or including the intercept term in X.

The effect of collinearity can be seen by this expression for var $\hat{\beta}_j$:

$$\text{var } \hat{\beta}_j = \sigma^2 \left(\frac{1}{1 - R_j^2} \right) \frac{1}{\sum_i (x_{ij} - \bar{x}_j)^2}$$

We can see that if the predictor x_j does not vary much, then the variance of $\hat{\beta}_j$ will be large. If R_j^2 is close to one, then the *variance inflation factor* $(1 - R_j^2)^{-1}$ will be large and so var $\hat{\beta}_j$ will also be large.

This equation also tells us which designs will minimize the variance of the regression coefficients if we have the ability to choose the X. Orthogonality means that $R_j^2 = 0$ which minimizes the variance. Also we can maximize $S_{x_j x_j}$ by spreading X as much as possible. The maximum is attained by placing half the points at the minimum practical value and half at the maximum. Unfortunately, this design assumes the linearity of the effect and would make it impossible to check for any curvature. So, in practice, we might put some design points in the middle of the range to allow checking of the fit.

Car drivers like to adjust the seat position for their own comfort. Car designers would find it helpful to know where different drivers will position the seat depending on their size and age. Researchers at the HuMoSim laboratory at the University of Michigan collected data on 38 drivers. They measured age in years, weight in pounds, height with shoes and without shoes in centimeters, seated height arm length, thigh length, lower leg length and hipcenter, the horizontal distance of the midpoint of the hips from a fixed location in the car in millimeters. We fit a model with all the predictors:

```
import faraway.datasets.seatpos
seatpos = faraway.datasets.seatpos.load()
lmod = smf.ols(
    'hipcenter ~ Age+Weight+HtShoes+Ht+Seated+Arm+Thigh+Leg',
    seatpos).fit()
lmod.sumary()
```

	coefs	stderr	tvalues	pvalues
Intercept	436.432	166.572	2.62	0.0138
Age	0.776	0.570	1.36	0.1843
Weight	0.026	0.331	0.08	0.9372
HtShoes	-2.692	9.753	-0.28	0.7845
Ht	0.601	10.130	0.06	0.9531
Seated	0.534	3.762	0.14	0.8882
Arm	-1.328	3.900	-0.34	0.7359
Thigh	-1.143	2.660	-0.43	0.6706
Leg	-6.439	4.714	-1.37	0.1824

n=38 p=9 Residual SD=37.720 R-squared=0.69

This model already shows the signs of collinearity. The R^2 is not small, but none of the individual predictors is significant. We take a look at the pairwise correlations:

```
seatpos.iloc[:,:-1].corr().round(3)
```

	Age	Weight	HtShoes	Ht	Seated	Arm	Thigh	Leg
Age	1.000	0.081	-0.079	-0.090	-0.170	0.360	0.091	-0.042
Weight	0.081	1.000	0.828	0.829	0.776	0.698	0.573	0.784
HtShoes	-0.079	0.828	1.000	0.998	0.930	0.752	0.725	0.908
Ht	-0.090	0.829	0.998	1.000	0.928	0.752	0.735	0.910
Seated	-0.170	0.776	0.930	0.928	1.000	0.625	0.607	0.812
Arm	0.360	0.698	0.752	0.752	0.625	1.000	0.671	0.754
Thigh	0.091	0.573	0.725	0.735	0.607	0.671	1.000	0.650
Leg	-0.042	0.784	0.908	0.910	0.812	0.754	0.650	1.000

There are several large pairwise correlations between predictors. Now we check the eigendecomposition of $X^T X$ (not including the intercept in X):

```
X = lmod.model.wexog[:,1:]
XTX = X.T @ X
evals, evecs = np.linalg.eig(XTX)
evals = np.flip(np.sort(evals))
evals
array([3.65367136e+06, 2.14794802e+04, 9.04322529e+03, 2.98952599e+02,
       1.48394821e+02, 8.11739742e+01, 5.33619434e+01, 7.29820918e+00])
```

The function np.linalg.eig does not necessarily sort the eigenvalues in descending order — we have enforced that. Now we compute the condition numbers:

```
np.sqrt(evals[0]/evals[1:])
array([ 13.04226011,  20.10032434, 110.55122882, 156.91171478,
       212.15649705, 261.66697969, 707.5491072 ])
```

There is a wide range in the eigenvalues, and several condition numbers are large.

This means that problems are being caused by more than just one linear combination. Now check the variance inflation factors (VIFs). For the first variable this is:

```
from patsy import dmatrix
X = dmatrix("Age+Weight+HtShoes+Ht+Seated+Arm+Thigh+Leg",
    seatpos, return_type='dataframe')
lmod = sm.OLS(X['Age'],X.drop('Age',axis=1)).fit()
lmod.rsquared, 1/(1-lmod.rsquared)
(0.4995, 1.9979)
```

which is moderate in size — the VIF for orthogonal predictors is one. Now we compute all the VIFs:

```
from statsmodels.stats.outliers_influence \
    import variance_inflation_factor
vif = [variance_inflation_factor(X.values, i) \
    for i in range(X.shape[1])]
pd.Series(vif, X.columns)
Intercept   741.029693
Age           1.997931
Weight        3.647030
HtShoes     307.429378
Ht          333.137832
Seated        8.951054
Arm           4.496368
Thigh         2.762886
Leg           6.694291
```

The intercept value should always be ignored since it makes little sense for a VIF. There is much variance inflation. For example, we can interpret $\sqrt{307.4} = 17.5$ as telling us that the standard error for height with shoes is 17.5 times larger than it would have been without collinearity. We cannot apply this as a correction because we did not actually observe orthogonal data, but it does give us a sense of the size of the effect.

There is substantial instability in these estimates. Measuring the hipcenter is difficult to do accurately and we can expect some variation in these values. Suppose the measurement error had a SD of 10 mm. Let's see what happens when we add a random perturbation of this size to the response:

```
seatpos['hiperb'] = seatpos.hipcenter+ \
    np.random.normal(scale=10,size=38)
lmod = smf.ols(
    'hipcenter ~ Age+Weight+HtShoes+Ht+Seated+Arm+Thigh+Leg',
    seatpos).fit()
lmodp = smf.ols(
    'hiperb ~ Age+Weight+HtShoes+Ht+Seated+Arm+Thigh+Leg',
    seatpos).fit()
pd.DataFrame([lmod.params, lmodp.params],
    index=['original','perturbed']).round(3)
```

	Intercept	Age	Weight	HtShoes	Ht	Seated	Arm	Thigh	Leg
original	436.432	0.776	0.026	-2.692	0.601	0.534	-1.328	-1.143	-6.439
perturbed	430.322	0.746	0.067	-4.824	2.636	0.857	-1.393	-1.860	-5.779

We see large changes in some of the coefficients, indicating their sensitivity to the response values caused by the collinearity. We compare the R^2's for the two models:

```
lmod.rsquared, lmodp.rsquared
(0.687, 0.669)
```

We see that these are similar.

We have too many variables that are trying to do the same job of explaining the response. We can reduce the collinearity by carefully removing some of the variables. We can then make a more stable estimation of the coefficients and come to a more secure conclusion regarding the effect of the remaining predictors on the response. But we should not conclude that the variables we drop have nothing to do with the response.

Examine the full correlation matrix above. Consider just the correlations of the length variables:

```
pd.DataFrame.corr(X.iloc[3:,3:]).round(3)
         HtShoes Ht      Seated  Arm     Thigh   Leg
HtShoes  1.000   0.998   0.930   0.722   0.710   0.896
Ht       0.998   1.000   0.929   0.724   0.720   0.898
Seated   0.930   0.929   1.000   0.603   0.576   0.803
Arm      0.722   0.724   0.603   1.000   0.670   0.723
Thigh    0.710   0.720   0.576   0.670   1.000   0.626
Leg      0.896   0.898   0.803   0.723   0.626   1.000
```

These six variables are strongly correlated with each other — any one of them might do a good job of representing the other. We pick height as the simplest to measure. We are not claiming that the other predictors are not associated with the response, just that we do not need them all to predict the response:

```
smf.ols('hipcenter ~ Age+Weight+Ht', seatpos).fit().sumary()
           coefs    stderr  tvalues pvalues
Intercept  528.298  135.313 3.90    0.0004
Age        0.520    0.408   1.27    0.2116
Weight     0.004    0.312   0.01    0.9891
Ht         -4.212   0.999   -4.22   0.0002
```

n=38 p=4 Residual SD=36.486 R-squared=0.66

Comparing this with the original fit, we see that the fit is very similar in terms of R^2, but many fewer predictors are used. Further simplification is clearly possible.

If you want to keep all your variables in the model, you should consider alternative methods of estimation such as ridge regression as described in Section 11.3.

The effect of collinearity on prediction is less serious. The accuracy of the prediction depends on where the prediction is to be made. The greater the distance is from the observed data, the more unstable the prediction. Of course, this is true for all data but collinear data covers a much smaller fraction of the predictor space than it might first appear. This means that predictions tend to be greater extrapolations than with data that are closer to the orthogonality.

Exercises

1. Using the `faithful` data, fit a regression of `duration` on `waiting`. Assuming that there was a measurement error in `waiting` with an SD of 30 seconds, use the SIMEX method to obtain a better estimate of the slope. You can obtain the data with:

```
gd = sm.datasets.get_rdataset("faithful", "datasets")
faithful = gd.data
```

and read the documentation with:

```
print(gd.__doc__)
```

2. What would happen if the SIMEX method was applied to the response error variance rather than predictor measurement error variance?

3. Using the `divusa` data:

   (a) Fit a regression model with `divorce` as the response and `unemployed`, `femlab`, `marriage`, `birth` and `military` as predictors. Compute the condition numbers and interpret their meanings.

   (b) For the same model, compute the VIFs. Is there evidence that collinearity causes some predictors not to be significant? Explain.

   (c) Does the removal of insignificant predictors from the model reduce the collinearity? Investigate.

4. You can obtain the `longley` data from the `statsmodels` package with:

   ```
   longley = sm.datasets.longley.load_pandas().data
   ```

   Fit a model with `TOTEMP` as the response and the other variables as predictors.

   (a) Compute and comment on the condition numbers.

   (b) Compute and comment on the correlations between the predictors.

   (c) Compute the variance inflation factors.

5. For the `prostate` data, fit a model with `lpsa` as the response and the other variables as predictors.

   (a) Compute and comment on the condition numbers.

   (b) Compute and comment on the correlations between the predictors.

   (c) Compute the variance inflation factors.

6. Using the `cheddar` data, fit a linear model with `taste` as the response and the other three variables as predictors.

   (a) Is the predictor `Lactic` statistically significant in this model?

   (b) Give the Python command to extract the p-value for the test of $\beta_{lactic} = 0$.

   (c) Add normally distributed errors to `Lactic` with mean zero and standard deviation 0.01 and refit the model. Now what is the p-value for the previous test?

   (d) Repeat this same calculation of adding errors to `Lactic` 1000 times within for loop. Save the p-values into a vector. Report on the average p-value. Does this much measurement error make a qualitative difference to the conclusions?

   (e) Repeat the previous question but with a standard deviation of 0.1. Does this much measurement error make an important difference?

7. Use the `mba` dataset with `happy` as the response and the other variables as predictors. Discuss standardization of the variables with the aim of helping the interpretation of the model fit.

8. Use the `fat` data, fitting the model described in Section 4.2.

(a) Compute the condition numbers and variance inflation factors. Comment on the degree of collinearity observed in the data.

(b) Cases 39 and 42 are unusual. Refit the model without these two cases and recompute the collinearity diagnostics. Comment on the differences observed from the full data fit.

(c) Fit a model with brozek as the response and just age, weight and height as predictors. Compute the collinearity diagnostics and compare to the full data fit.

(d) Compute a 95% prediction interval for brozek for the median values of age, weight and height.

(e) Compute a 95% prediction interval for brozek for age=40, weight=200 and height=73. How does the interval compare to the previous prediction?

(f) Compute a 95% prediction interval for brozek for age=40, weight=130 and height=73. Are the values of predictors unusual? Comment on how the interval compares to the previous two answers.

Chapter 8

Problems with the Error

We have assumed that the error ε is independent and identically distributed (i.i.d.). Furthermore, we have also assumed that the errors are normally distributed in order to carry out the usual statistical inference. We have seen that these assumptions can often be violated and we must then consider alternatives. When the errors are dependent, we can use *generalized least squares* (GLS). When the errors are independent, but not identically distributed, we can use *weighted least squares* (WLS), which is a special case of GLS. Sometimes, we have a good idea how large the error should be, but the residuals may be much larger than we expect. This is evidence of a *lack of fit*. When the errors are not normally distributed, we can use *robust regression*.

8.1 Generalized Least Squares

Until now we have assumed that var $\varepsilon = \sigma^2 I$, but sometimes the errors have non-constant variance or are correlated. Suppose instead that var $\varepsilon = \sigma^2 \Sigma$ where σ^2 is unknown but Σ is known — in other words, we know the correlation and relative variance between the errors, but we do not know the absolute scale of the variation. For now, it might seem redundant to distinguish between σ and Σ, but we will see how this will be useful later.

We can write $\Sigma = SS^T$, where S is a triangular matrix using the Choleski decomposition, which be can be viewed as a square root for a matrix. We can transform the regression model as follows:

$$
\begin{aligned}
y &= X\beta + \varepsilon \\
S^{-1}y &= S^{-1}X\beta + S^{-1}\varepsilon \\
y' &= X'\beta + \varepsilon'
\end{aligned}
$$

Now we find that:

$$
\text{var } \varepsilon' = \text{var } (S^{-1}\varepsilon) = S^{-1}(\text{var } \varepsilon)S^{-T} = S^{-1}\sigma^2 SS^T S^{-T} = \sigma^2 I
$$

So we can reduce GLS to ordinary least squares (OLS) by a regression of $y' = S^{-1}y$ on $X' = S^{-1}X$ which has error $\varepsilon' = S^{-1}\varepsilon$ that is i.i.d. We have transformed the problem to the standard case. In this transformed model, the sum of squares is:

$$
(S^{-1}y - S^{-1}X\beta)^T(S^{-1}y - S^{-1}X\beta) = (y - X\beta)^T S^{-T}S^{-1}(y - X\beta) = (y - X\beta)^T\Sigma^{-1}(y - X\beta)
$$

which is minimized by:

$$
\hat{\beta} = (X^T\Sigma^{-1}X)^{-1}X^T\Sigma^{-1}y
$$

115

We find that:

$$\text{var } \hat{\beta} = (X^T \Sigma^{-1} X)^{-1} \sigma^2$$

Since $\varepsilon' = S^{-1}\varepsilon$, diagnostics should be applied to the residuals, $S^{-1}\hat{\varepsilon}$. If we have the right Σ, then these should be approximately i.i.d.

The main problem in applying GLS in practice is that Σ may not be known and we have to estimate it. We start by loading the packages we will use:

```
import pandas as pd
import numpy as np
import matplotlib.pyplot as plt
import scipy as sp
import statsmodels.api as sm
import statsmodels.formula.api as smf
import seaborn as sns
import faraway.utils
```

Let's take another look at the global warming data first considered in Section 6.1.3. We found evidence of serial correlation by looking at successive residuals. We demonstrate how we can model this using GLS. We start with the OLS solution by fitting a linear model:

```
import faraway.datasets.globwarm
globwarm = faraway.datasets.globwarm.load()
lmod = smf.ols('nhtemp ~ wusa + jasper + westgreen + chesapeake +\
    tornetrask +  urals + mongolia + tasman', globwarm).fit()
lmod.sumary()
```

```
              coefs stderr tvalues pvalues
Intercept    -0.243  0.027   -8.98  0.0000
wusa          0.077  0.043    1.80  0.0736
jasper       -0.229  0.078   -2.93  0.0040
westgreen     0.010  0.042    0.23  0.8192
chesapeake   -0.032  0.034   -0.94  0.3473
tornetrask    0.093  0.045    2.06  0.0416
urals         0.185  0.091    2.03  0.0446
mongolia      0.042  0.046    0.92  0.3610
tasman        0.115  0.030    3.83  0.0002
```

```
n=145 p=9 Residual SD=0.176 R-squared=0.48
```

In data collected over time such as this, successive errors could be correlated. We have calculated this by computing the correlation between the vector of residuals with the first and then the last term omitted.

```
np.corrcoef(lmod.resid.iloc[:-1],lmod.resid.iloc[1:]).round(3)
array([[1.   , 0.583],
       [0.583, 1.   ]])
```

We can see a correlation of 0.583 between successive residuals. A more direct way to achieve this is:

```
lmod.resid.autocorr()
0.583
```

The simplest way to model this is the autoregressive form:

$$\varepsilon_{i+1} = \phi \varepsilon_i + \delta_i$$

where $\delta_i \sim N(0, \tau^2)$. We use the GLSAR() function from statsmodels to fit this model:

```
X = lmod.model.wexog
y = lmod.model.wendog
gmod = sm.GLSAR(y, X, rho=1)
res=gmod.iterative_fit(maxiter=6)
gmod.rho.round(3)
array([0.582])
```

We see that the estimated value of ρ is very similar to the OLS residual calculation. We examine the model summary:

```
res.summary().tables[1]
```

	coef	std err	t	P>|t|	[0.025	0.975]
const	-0.2315	0.050	-4.658	0.000	-0.330	-0.133
x1	0.0634	0.077	0.823	0.412	-0.089	0.216
x2	-0.2120	0.143	-1.488	0.139	-0.494	0.070
x3	0.0066	0.072	0.092	0.927	-0.136	0.149
x4	-0.0144	0.058	-0.247	0.805	-0.130	0.101
x5	0.0577	0.076	0.760	0.448	-0.092	0.208
x6	0.2321	0.170	1.366	0.174	-0.104	0.568
x7	0.0490	0.082	0.596	0.552	-0.114	0.212
x8	0.1249	0.055	2.266	0.025	0.016	0.234

The standard errors of $\hat{\beta}$ are much larger, and only one of the predictors is statistically significant in the output. However, there is substantially collinearity between the predictors so this should not be interpreted as "no predictor effect". Also you should understand that correlation between the predictors and correlation between the errors are different phenomena and there is no necessary link between the two. It is not surprising to see significant autocorrelation in this example because the proxies can only partially predict the temperature and we would naturally expect some carryover effect from one year to the next.

Another situation where correlation between errors might be anticipated is where observations are grouped in some way. Other examples where correlated errors can arise are in spatial data where the relative locations of the observations can be used to model the error correlation. In other cases, one may suspect a correlation between errors but have no structure to suggest a parameterized form such as serial correlation or compound symmetry. The problem is that there are too many pairwise correlations to be estimated and not enough data to do it.

8.2 Weighted Least Squares

Sometimes the errors are uncorrelated, but have unequal variance where the form of the inequality is known. In such cases, Σ is diagonal but the entries are not equal. Weighted least squares (WLS) is a special case of GLS and can be used in this situation. We set $\Sigma = \text{diag}(1/w_1, \ldots, 1/w_n)$, where the w_i are the *weights* so $S = \text{diag}(\sqrt{1/w_1}, \ldots, \sqrt{1/w_n})$. We then regress $\sqrt{w_i}y_i$ on $\sqrt{w_i}x_i$ (although the column of ones in the X-matrix needs to be replaced with $\sqrt{w_i}$). When weights are used, the residuals must be modified to use $\sqrt{w_i}\hat{\varepsilon}_i$. We see that cases with low variability get a high weight and those with high variability a low weight. Some examples:

1. Errors proportional to a predictor: $\text{var}(\varepsilon_i) \propto x_i$ suggests $w_i = x_i^{-1}$. One might

choose this option after observing a positive relationship in a plot of $|\hat{\varepsilon}_i|$ against x_i.

2. When the Y_i are the averages of n_i observations, then $\text{var } y_i = \text{var } \varepsilon_i = \sigma^2/n_i$, which suggests $w_i = n_i$. Responses that are averages arise quite commonly, but take care that the variance in the response really is proportional to the group size. For example, consider the life expectancies for different countries. At first glance, one might consider setting the weights equal to the populations of the countries, but notice that there are many other sources of variation in life expectancy that would dwarf the population size effect.

3. When the observed responses are known to be of varying quality, weights may be assigned $w_i = 1/\text{var}(y_i)$.

Elections for the French presidency proceed in two rounds. In 1981, there were 10 candidates in the first round. The top two candidates then went on to the second round, which was won by François Mitterand over Valéry Giscard-d'Estaing. The losers in the first round can gain political favors by urging their supporters to vote for one of the two finalists. Since voting is private, we cannot know how these votes were transferred; we might hope to infer from the published vote totals how this might have happened. Anderson and Loynes (1987) published data on these vote totals in every fourth department of France:

```
import faraway.datasets.fpe
fpe = faraway.datasets.fpe.load()
fpe.head()
```

	EI	A	B	... K	A2	B2	N
Ain	260	51	64	... 3	105	114	17
Alpes	75	14	17	... 1	32	31	5
Ariege	107	27	18	... 1	57	33	6
Bouches.du.Rhone	1036	191	204	... 6	466	364	30
Charente.Maritime	367	71	76	... 2	163	142	17

A and B stand for Mitterand's and Giscard's votes in the first round, respectively, while A2 and B2 represent their votes in the second round. C–K are the first round votes of the other candidates while EI denotes *electeur inscrits* or registered voters. All numbers are in thousands. The total number of voters in the second round was greater than the first — we can compute the difference as N. We will treat this group like another first round candidate (although we could reasonably handle this differently).

Now we can represent the transfer of votes as:

$$A2 = \beta_A A + \beta_B B + \beta_C C + \beta_D D + \beta_E E + \beta_F F + \beta_G G + \beta_H H + \beta_J J + \beta_K K + \beta_N N$$

where β_i represents the proportion of votes transferred from candidate i to Mitterand in the second round. We can equally well do this for Giscard-d'Estaing but then the β's will simply be the remaining proportions so it's not necessary to do both. Our first model is suggested by this equation, where we use -1 in the model formula because there is no intercept.

```
lmod = smf.wls("A2 ~ A + B + C + D + E + F + G + H + J + K +N -1",
    fpe).fit()
```

We would expect the transfer proportions to vary somewhat between departments, so if we treat the above as a regression equation, there will be some error from department to department. The error will have a variance in proportion to the number of voters because it will be like a variance of a sum rather than a mean. Since the weights should be inversely proportional to the variance, this suggests that the weights should be set to 1/EI. We fit a weighted least squares model:

```
wmod = smf.wls("A2 ~ A + B + C + D + E + F + G + H + J + K +N -1",
      fpe, weights = 1/fpe.EI ).fit()
```

Only the relative proportions of the weights matter — for example, suppose we multiply the weights by an arbitray 53:

```
wmod53 = smf.wls("A2 ~ A + B + C + D + E + F + G + H + J + K+N-1",
      fpe, weights = 53/fpe.EI ).fit()
```

Now let's examine the coefficients from these three models:

```
pd.DataFrame([lmod.params, wmod.params, wmod53.params],
      index=['no weights','weights','weights*53']).round(3)
```

	A	B	C	D	E	F	G
no weights	1.075	-0.125	0.257	0.905	0.671	0.783	2.166
weights	1.067	-0.105	0.246	0.926	0.249	0.755	1.972
weights*53	1.067	-0.105	0.246	0.926	0.249	0.755	1.972

H	J	K	N
-0.854	0.144	0.518	0.558
-0.566	0.612	1.211	0.529
-0.566	0.612	1.211	0.529

We see that using weights makes a difference but only the relative size of the weights matters.

Now there is one remaining difficulty, unrelated to the weighting, in that proportions are supposed to be between zero and one. We can impose an *ad hoc* fix by truncating the coefficients that violate this restriction either to zero or one as appropriate. We achieve this by modifying the response (using variables with fixed coefficient one) and omitting variables that have fixed coefficient zero:

```
y = fpe.A2 - fpe.A - fpe.G - fpe.K
X = fpe.loc[:,["C","D","E","F","N"]]
wmod = sm.WLS(y, X, weights = 1/fpe.EI ).fit()
wmod.params
C   0.225773
D   0.969977
E   0.390204
F   0.744240
N   0.608539
```

We see that voters for the Communist candidate D apparently almost all voted for the Socialist Mitterand in the second round. However, we see that around 20% of the voters for the Gaullist candidate C voted for Mitterand. This is surprising since these voters would normally favor the more right wing candidate, Giscard. This appears to be the decisive factor. We see that of the larger blocks of smaller candidates, the Ecology party voters, E, roughly split their votes as did the first round non-voters. The other candidates had very few voters, and so their behavior is less interesting.

This analysis is somewhat crude, and more sophisticated approaches are discussed in Anderson and Loynes (1987). We bake the weights into the variables first:

```
y = fpe.A2
```

```
X = fpe.loc[:,["A","B","C","D","E","F","G","H","J","K","N"]]
weights = 1/fpe.EI
Xw = (X.T * np.sqrt(weights)).T
yw = y * np.sqrt(weights)
```

and now using a constrained optimization:

```
res = sp.optimize.lsq_linear(Xw, yw, bounds=(0, 1))
pd.Series(np.round(res.x,3),index=lmod.params.index)
```

```
A    1.000
B    0.000
C    0.208
D    0.969
E    0.359
F    0.743
G    1.000
H    0.367
J    0.000
K    1.000
N    0.575
```

The results are quite similar for the candidates C, D, E and N who have substantial numbers of votes, but the coefficients for small party candidates vary much more.

In the previous example, the weights were determined by the structure of the data. In some cases, it is clear that the variance of the response is not constant but there is no clear message about the form of this variation. Consider some data on the speed and stopping distances of cars in the 1920s.

```
import faraway.datasets.cars
cars = faraway.datasets.cars.load()
sns.regplot(x='speed', y = 'dist', data = cars)
```

The seaborn package provides the function regplot which conveniently plots the linear model fit on top of the data as seen in the first panel of Figure 8.1. We can see that the variance of the response about the line increases with the predictor. We might wish to use weights to allow for this, but what form should they take? A crude estimate of the variance for the ith point is given by $\hat{e}^2$. We might model the relationship between $\hat{y}$ (or just x in this example) and $\hat{e}^2$. We should then set the weights to be inversely proportional to the estimated variance from this model.

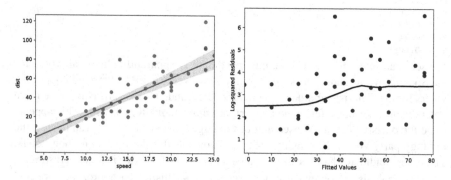

Figure 8.1   Linear fit to the cars data is shown on the left. An estimation of the form of the non-constant variance is shown on the right.

Instead of modeling $\hat{\varepsilon}^2$, it is better to model $\log(\hat{\varepsilon}^2)$ for two reasons. One is that squared residuals can sometimes be quite large and influential. The second is that some models may predict a negative variance for some inputs. As we need to exponentiate to get back to the original scale, this problem is avoided. Let's see how this works for our example:

```
lmod = smf.ols('dist ~ speed', cars).fit()
g = sns.regplot(lmod.fittedvalues, np.sqrt(abs(lmod.resid)),
    lowess = True)
g.set_xlabel("Fitted Values")
g.set_ylabel("Log-squared Residuals")
```

Instead of showing a fitted line, we have used the *Lowess* method to fit a smooth curve to the plot as seen in the second panel of Figure 8.1. We see that the variance is increasing. We could just fit a line here, but we can use the lowess fit to compute suitable weights:

```
x = lmod.fittedvalues
y = np.log(lmod.resid**2)
z = sm.nonparametric.lowess(y,x)
w = 1.0/np.exp(z[:,1])
wmod = smf.wls('dist ~ speed', cars, weights =w).fit()
wmod.summary().tables[1]
```

	coef	std err	t	P>|t|	[0.025	0.975]
Intercept	-14.8498	4.876	-3.045	0.004	-24.655	-5.045
speed	3.7638	0.348	10.819	0.000	3.064	4.463

The lowess function is found in the statsmodels package and was used by the seaborn plot. We are simply reconstructing what was computed in the plot. The lowess function returns a two-column matrix with the first column being the predictor values. We want the response values found in the second column. We square these and supply these as weights. We examine only the coefficients and associated values. We should compare these with the corresponding OLS values:

```
print(lmod.summary().tables[1])
```

	coef	std err	t	P>|t|	[0.025	0.975]
Intercept	-17.5791	6.758	-2.601	0.012	-31.168	-3.990
speed	3.9324	0.416	9.464	0.000	3.097	4.768

We see that the results are different although not substantially. The use of weights will only make much difference when the variance is much more variable. In a mild cases such as this, it does not make much difference. This suggests that when viewing diagnostic plots, action is necessary only when the non-constant variance is strong.

8.3   Testing for Lack of Fit

How can we tell whether a model fits the data? If the model is correct, then $\hat{\sigma}^2$ should be an unbiased estimate of σ^2. If we have a model that is not complex enough to fit the data or simply takes the wrong form, then $\hat{\sigma}^2$ will tend to overestimate σ^2. The situation is illustrated in Figure 8.2 where the residuals from the incorrect constant fit will lead to an overestimate of σ^2. Alternatively, if our model is too complex and overfits the data, then $\hat{\sigma}^2$ will be an underestimate.

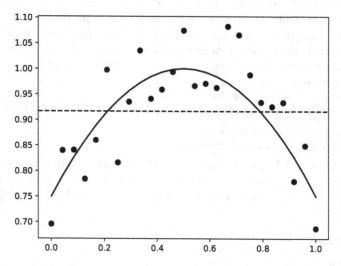

Figure 8.2   True quadratic fit shown with the solid line and incorrect linear fit shown with the dotted line. Estimate of σ^2 will be unbiased for the quadratic model, but too large for the linear model.

This suggests a possible testing procedure — we should compare $\hat{\sigma}^2$ to σ^2. But this requires us to know σ^2. In a few cases, we might actually know σ^2 – for example, when measurement error is the only source of variation and we know its variance because we are very familiar with the measurement device. This is rather uncommon because we usually have more than just measurement error.

We might compare the $\hat{\sigma}^2$ from our chosen regression model to one obtained from another model. But this may just be the same as the F-test introduced in Chapter 3. This would indicate a preference between the two models but could not tell us that the preferred model fit the data. We need to make a comparison to some model-free estimate of σ^2.

We can do this if we have repeated values of the response for one or more fixed values of x. These replicates do need to be truly independent. They cannot just be repeated measurements on the same subject or unit. For example, the cases in the data may be people and the response might be blood pressure. We might sensibly repeat these measurements of blood pressure. But such repeated measures would only reveal the within-subject variability or the measurement error. We need different people but with the same predictor values. This would give us the between-subject variability and allow us to construct an estimate of σ^2 that does not depend on a particular model.

Let y_{ij} be the ith observation in the group of true replicates j. The "pure error" or model-free estimate of σ^2 is given by SS_{pe}/df_{pe} where:

$$SS_{pe} = \sum_{j} \sum_{i} (y_{ij} - \bar{y}_j)^2$$

where $\bar{y}_j$ is the mean within replicate group j. The degrees of freedom are $df_{pe} = \sum_j (\#replicates_j - 1) = n - \#groups$.

There is a convenient way to compute the estimate. Fit a model that assigns one parameter to each group of observations with fixed x, then the $\hat{\sigma}^2$ from this model will be the pure error $\hat{\sigma}^2$. This model is saturated and tells us nothing interesting about the relationship between the predictors and the response. It simply fits a mean to each group of replicates. Values of x with no replication will be fit exactly and will not contribute to the estimate of σ^2. Comparing this model to the regression model using the standard F-test gives us the lack-of-fit test.

The data for this example consist of 13 specimens of 90/10 Cu–Ni alloys with varying percentages of iron content. The specimens were submerged in seawater for 60 days and the weight loss due to corrosion was recorded in units of milligrams per square decimeter per day. The data come from Draper and Smith (1998). We fit a straight-line model: We load in and plot the data, as seen in Figure 8.3:

```
import faraway.datasets.corrosion
corrosion = faraway.datasets.corrosion.load()
lmod = smf.ols('loss ~ Fe', corrosion).fit()
lmod.sumary()
          coefs stderr tvalues pvalues
Intercept 129.787  1.403   92.52  0.0000
Fe         -24.020  1.280  -18.77  0.0000

n=13 p=2 Residual SD=3.058 R-squared=0.97
```

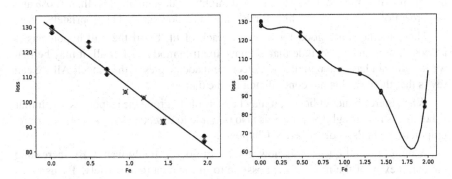

Figure 8.3   Linear fit to the Cu–Ni corrosion data is shown on the left. Group means are denoted by crosses. A polynomial fit to the data is shown on the right.

We have an R^2 of 97% and an apparently good fit to the data. We now fit a model that reserves a parameter for each group of data with the same value of x. This is accomplished by declaring the predictor to be a factor. We will describe this in more detail in Chapter 15:

```
corrosion['Fefac'] = corrosion['Fe'].astype('category')
amod = smf.ols('loss ~ Fefac', corrosion).fit()
```

The fitted values are the means in each group. We put the data, linear fit and mean value for each value of the predictor on the same plot:

```
fig, ax = plt.subplots()
```

```
ax.scatter(corrosion.Fe, corrosion.loss)
plt.xlabel("Fe")
plt.ylabel("loss")
xr = np.array(ax.get_xlim())
ax.plot(xr, lmod.params[0] + lmod.params[1] * xr)
ax.scatter(corrosion.Fe, amod.fittedvalues, marker='x',s=100)
```

We can now compare the two models in the usual way:

```
amod.compare_f_test(lmod)
```

```
(9.28, 0.00862, 5.0)
```

The low p-value indicates that we must conclude that there is a lack of fit. The reason is that the pure error SD $\sqrt{(11.8/6)} = 1.4$, is substantially less than the regression standard error of 3.06. We might investigate models other than a straight line although no obvious alternative is suggested by the plot. Before considering other models, we would first find out whether the replicates are genuine. Perhaps the low pure error SD can be explained by some correlation in the measurements. They may not be genuine replicates. Another possible explanation is that an unmeasured third variable is causing the lack of fit.

When there are replicates, it is impossible to get a perfect fit. Even when there is a parameter assigned to each group of x-values, the residual sum of squares will not be zero. For the factor-based model above, the R^2 is 99.7%. So even this saturated model does not attain a 100% value for R^2. For these data, it is a small difference but in other cases, the difference can be substantial. In these cases, one should realize that the maximum R^2 that may be attained might be substantially less than 100% and so perceptions about a good value for R^2 should be downgraded appropriately.

These methods are good for detecting lack of fit, but if the null hypothesis is accepted, we cannot conclude that we have the true model. After all, it may be that we just did not have enough data to detect the inadequacies of the model. All we can say is that the model is not contradicted by the data.

When there are no replicates, it may be possible to group the responses for similar x, but this is not straightforward. It is also possible to detect lack of fit by less formal, graphical methods as described in Chapter 6.

A more general question is how good a fit do you really want? By increasing the complexity of the model, it is possible to fit the data more closely. By using as many parameters as data points, we can fit the data exactly. Very little is achieved by doing this since we learn nothing beyond the data and any predictions made using such a model will tend to have high variance. The question of how complex a model to fit is difficult and fundamental. For example, we can fit the mean responses for the previous example exactly using a sixth order polynomial:

```
pc = np.polyfit(corrosion.Fe, corrosion.loss, 6)
```

Now look at this fit:

```
fig, ax = plt.subplots()
ax.scatter(corrosion.Fe, corrosion.loss)
plt.xlabel("Fe")
plt.ylabel("loss")
ax.scatter(corrosion.Fe, amod.fittedvalues, marker='x')
grid = np.linspace(0,2,50)
ax.plot(grid,np.poly1d(pc)(grid))
```

as shown in the right panel of Figure 8.3. The fit of this model is excellent — for example:
```
pc, rss, _, _, _ = np.polyfit(corrosion.Fe, corrosion.loss,
    6, full=True)
tss = np.sum((corrosion.loss-np.mean(corrosion.loss))**2)
1-rss/tss
array([0.997])
```
but it is clearly ridiculous. There is no plausible reason corrosion loss should suddenly drop at 1.7 and thereafter increase rapidly. This is a consequence of overfitting the data. This illustrates the need not to become too focused on measures of fit like R^2. The fit needs to reflect knowledge of the subject matter and simplicity in modeling is a virtue.

8.4 Robust Regression

When the errors are normally distributed, least squares regression is best. But when the errors follow some other distribution, other methods of model fitting may be considered. Short-tailed errors are not so much of a problem, but long-tailed error distributions can cause difficulties because a few extreme cases can have a large effect on the fitted model.

It is important to think about the cause of these extreme values. In some cases, they are mistakes and we just want some way of automatically excluding them from the analysis. In other cases, the extreme values really were observed and are part of the process we are trying to model. Robust regression is designed to estimate the mean relationship between the predictors and response, $EY = X\beta$. It doesn't worry about where the outliers come from, but the analyst does need to think about this.

We developed methods of detecting outliers in Section 6.2.2. We could use these methods to remove the largest residuals as outliers and then just use least squares. This does not work very well if there are multiple outliers, as such points can mutually influence the fit and effectively hide their presence. Furthermore, outlier rejection-based methods tend not to be statistically efficient for the estimation of β. Robust regression works better if you are dealing with more than one or two outliers.

In this section, we present two popular types of robust regression.

8.4.1 M-Estimation

M-estimates modify the least squares idea to choose β to minimize:

$$\sum_{i=1}^{n} \rho(y_i - x_i^T \beta)$$

Some possible choices among many for ρ are:

1. $\rho(x) = x^2$ is simply least squares.
2. $\rho(x) = |x|$ is called least absolute deviation (LAD) regression or L_1 regression.
3.

$$\rho(x) = \begin{cases} x^2/2 & \text{if } |x| \leq c \\ c|x| - c^2/2 & \text{otherwise} \end{cases}$$

is called Huber's method and is a compromise between least squares and LAD regression. c should be a robust estimate of σ. A value proportional to the median of $|\hat{e}|$ is suitable.

M-estimation is related to weighted least squares. The normal equations tell us that:

$$X^T(y - X\hat{\beta}) = 0$$

With weights and in nonmatrix form, this becomes:

$$\sum_{i=1}^{n} w_i x_{ij}(y_i - \sum_{j=1}^{p} x_{ij}\beta_j) = 0 \quad j = 1, \ldots p$$

Now differentiating the M-estimate criterion with respect to β_j and setting to zero we get:

$$\sum_{i=1}^{n} \rho'(y_i - \sum_{j=1}^{p} x_{ij}\beta_j)x_{ij} = 0 \quad j = 1, \ldots p$$

Now let $u_i = y_i - \sum_{j=1}^{p} x_{ij}\beta_j$ to get:

$$\sum_{i=1}^{n} \frac{\rho'(u_i)}{u_i} x_{ij}(y_i - \sum_{j=1}^{p} x_{ij}\beta_j) = 0 \quad j = 1, \ldots p$$

so we can make the identification of a weight function as $w(u) = \rho'(u)/u$. We find for our choices of ρ above that:

1. LS: $w(u)$ is constant and the estimator is simply ordinary least squares.

2. LAD: $w(u) = 1/|u|$. We see how the weight goes down as u moves away from zero so that more extreme observations get downweighted. Unfortunately, there is an asymptote at zero. This makes a weighting approach to fitting an LAD regression infeasible without some modification.

3. Huber:

$$w(u) = \begin{cases} 1 & \text{if } |u| \leq c \\ c/|u| & \text{otherwise} \end{cases}$$

We can see that this sensibly combines the downweighting of extreme cases with equal weighting for the middle cases.

There are many other choices for ρ that have been proposed. Computing an M-estimate requires some iteration because the weights depend on the residuals. The fitting methods alternate between fitting a WLS and recomputing the weights based on the residuals until convergence. We can get standard errors via WLS by $\hat{\text{var}}\,\hat{\beta} = \hat{\sigma}^2(X^T W X)^{-1}$ but we need to use a robust estimate of σ^2.

We demonstrate the methods on the Galápagos Islands data. Using least squares first:

```
import faraway.datasets.galapagos
galapagos = faraway.datasets.galapagos.load()
lsmod = smf.ols(
    'Species ~ Area + Elevation + Nearest + Scruz  + Adjacent',
    galapagos).fit()
lsmod.sumary()
```

	coefs	stderr	tvalues	pvalues
Intercept	7.068	19.154	0.37	0.7154
Area	-0.024	0.022	-1.07	0.2963
Elevation	0.319	0.054	5.95	0.0000
Nearest	0.009	1.054	0.01	0.9932
Scruz	-0.241	0.215	-1.12	0.2752
Adjacent	-0.075	0.018	-4.23	0.0003

n=30 p=6 Residual SD=60.975 R-squared=0.77

Least squares works well when there are normal errors, but performs poorly for long-tailed errors. Fit the robust regression using default from the RLM() function of statsmodels which is Huber T:

```
X = lsmod.model.wexog
y = lsmod.model.wendog
rlmod = sm.RLM(y,X).fit()
rlmod.summary()
```

	coef	std err	z	P>|z|
const	6.3626	12.366	0.515	0.607
Area	-0.0061	0.014	-0.422	0.673
Elevation	0.2476	0.035	7.146	0.000
Nearest	0.3590	0.681	0.528	0.598
Scruz	-0.1952	0.139	-1.404	0.160
Adjacent	-0.0546	0.011	-4.774	0.000

The R^2 statistic is not given because it does not make sense in the context of a robust regression. We see that the same two predictors, Elevation and Adjacent, are significant. The numerical values of the coefficients have changed somewhat and the standard errors are generally smaller.

It is worth looking at the weights assigned by the final fit. We extract and show those that are less than one. The remaining weights are all ones.

```
wts = rlmod.weights
wts[wts < 1]
```

Espanola	0.679642
Gardner1	0.661450
Gardner2	0.850097
Pinta	0.537700
SanCristobal	0.414224
SantaCruz	0.174601
SantaMaria	0.307863

We can see that a few islands are substantially discounted in the calculation of the robust fit. Provided we do not believe there are mistakes in the data for these cases, we should think carefully about what might be unusual about these islands.

The main purpose in analyzing these data is likely to explain the relationship between the predictors and the response. Although the robust fit gives numerically different output, the overall impression of what predictors are significant in explaining the response is unchanged. Thus the robust regression has provided some measure of confirmation. Furthermore, it has identified a few islands which are not fit so well by the model. If there had been more disagreement between the two sets of regression outputs, we would know which islands are responsible and deserve a closer look. If there is a substantial difference between the two fits, we find the robust one more trustworthy.

Robust regression is not a panacea. M-estimation does not address large leverage points. It also does not help us choose which predictors to include or what transformations of the variables to make. For this data, we have seen that a transformation of the response may be helpful which would completely change the robust fit. Hence robust methods are just part of the regression modeling toolkit and not a replacement.

We can also do LAD regression using the `QuantReg()` function. The default option does LAD while other options allow for quantile regression:

```
l1mod = sm.QuantReg(y,X).fit()
l1mod.summary()
```

	coef	std err	t	P>|t|
const	1.3145	16.754	0.078	0.938
Area	-0.0031	0.020	-0.156	0.877
Elevation	0.2321	0.047	4.945	0.000
Nearest	0.1637	0.922	0.177	0.861
Scruz	-0.1231	0.188	-0.654	0.520
Adjacent	-0.0519	0.015	-3.349	0.003

Again, there is some change in the coefficients but no real change in what appears significant.

For this example, we do not see any big qualitative difference in the coefficients and for want of evidence to the contrary, we might stick with least squares as the easiest to work with. Had we seen something different, we would need to find out the cause. Perhaps some group of observations were not being fit well and the robust regression excluded these points.

8.4.2   High Breakdown Estimators

Sometimes M-estimation can fail. Consider the star data presented in Section 6.2.2 as shown in Figure 8.4. We compute the least squares and default Huber fits:

```
import faraway.datasets.star
star = faraway.datasets.star.load()
gs1 = smf.ols('light ~ temp', star).fit()
X = gs1.model.wexog
gs2 = sm.RLM(star.light, X, data=star).fit()
gs3 = smf.ols('light ~ temp', star.loc[star.temp > 3.6,:]).fit()
plt.scatter(star.temp, star.light, label = None)
xr = np.array([min(star.temp), max(star.temp)])
plt.plot(xr, gs1.params[0] + gs1.params[1]*xr,'k-',label="OLS")
plt.plot(xr, gs2.params[0] + gs2.params[1]*xr,'k--',label="Huber")
plt.plot(xr, gs3.params[0] + gs3.params[1]*xr,'k:',label="OLS-4")
plt.legend()
```

The four stars on the right are atypical and we might wish the model to fit the rest of the data where light intensity increases with temperature. Unfortunately, the Huber estimator is unable to recognize the four stars as outliers to be ignored and is almost indistinguishable from the least squares estimator. We have added the least squares estimator which excludes the four stars from the data which is more like the fit we might hope for. In this example, we can plot the data and manually intervene to produce the desired result. In higher dimensions, such problems are harder to spot and in some applications, we might want the fit to proceed automatically without human supervision. An automated method to detect and exclude such points is wanted.

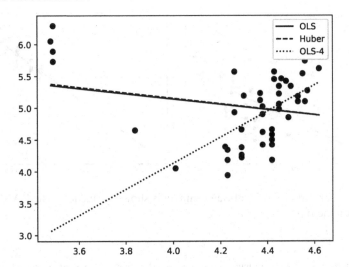

Figure 8.4   Regression fits compared. The OLS-4 label indicates the OLS fit where the four outliers are removed.

The *breakdown point* of an estimator is defined as the largest percentage of arbitrarily bad points that can be tolerated before the estimator fails. Sometimes the bad points are called *contaminated data*. *Resistant* estimators have higher breakdown points. The maximum possible value of the breakdown point is 50% since we cannot tolerate more than half the data being bad. In the star example, a relatively small proportion of contamination has caused the presumed robust Huber estimator to fail.

For univariate data, the median is known as a highly robust resistant estimator and has a breakdown point of 50%. One might suppose the LAD estimator is the equivalent of the median in the regression setting, but this fails on examples such as the star data. The Theil-Sen estimator is a better partner to the median. It can be computed using the following procedure.

1. Compute the slopes of the lines connecting all pairs of points in the data.

2. The estimated slope, $\hat{\beta}$ is the median of these slopes.

3. The estimated intercept is given by the median of the values $y_i - \hat{\beta}x_i$.

The procedure is available from the scikit-learn package:

```
from sklearn.linear_model import TheilSenRegressor
X = np.reshape(star.temp.array,(-1,1))
y = star.light
reg = TheilSenRegressor(random_state=0).fit(X,y)
plt.scatter(star.temp, star.light)
plt.plot(xr, reg.intercept_ + reg.coef_*xr,'k-')
```

The procedure requires that the predictor be supplied as a two-dimensional matrix rather than a vector, even though it has only one column in this example. This is because the procedure can be generalized to higher dimensional predictors by fitting planes rather than lines. The plot is shown in the first panel of Figure 8.5. We see that the estimator succeeds in capturing the trend in the bulk of the data. It is possible

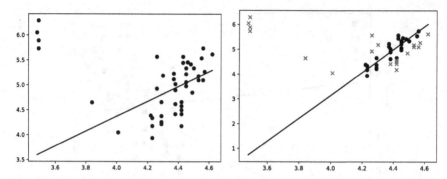

Figure 8.5   The fit of the Theil-Sen estimate is shown on the left. The RANSAC fit is shown on the right.

to compute a confidence interval using percentiles from the set of paired slopes, but this is not implemented in `TheilSenRegressor`.

The estimator has a breakdown point of 29.3% and so it is highly resistant. For larger samples and particularly in higher dimensions, it is necessary to economize on the computation by taking only a random sample of pairs of points.

The *Random Sample Consensus* (RANSAC) method is an algorithm developed in image processing that can be applied to regression problems. In this context, the procedure is:

1. Select a random sample of size just sufficient to estimate the parameters. For univariate regression, we would need only two points.

2. Compute the absolute values of the residuals. Residuals smaller than a threshold are called *inliers*.

We repeat this random sampling looking for the choice that maximizes the number of inliers. For small datasets, we might enumerate all possible subsets but for larger subsets, random sampling is necessary. Since we need to terminate the sampling to limit the computation, it is possible that we may not find the optimal solution. Also, one should realize that the outcome is somewhat random because of this sampling. A standard choice for the threshold is the median absolute deviation of the response values.

The procedure is implemented as RANSACRegressor in the scikit-learn package:

```
from sklearn.linear_model import   RANSACRegressor
reg = RANSACRegressor().fit(X,y)
i = reg.inlier_mask_
plt.scatter(star.temp[i], star.light[i])
plt.scatter(star.temp[~i], star.light[~i],marker='x')
plt.plot(xr, reg.estimator_.intercept_ + reg.estimator_.coef_*xr,
    'k-')
```

The fitted line can be seen in the second panel of Figure 8.5. The inliers are distinguished from the outliers. In this case, we see that many observations that are not that extreme are identified as "outliers". We could avoid this by choosing a higher threshold, but it can be difficult to choose this without the benefit of hindsight.

One major drawback to the RANSAC estimator is that it does not provide standard errors or other necessities of statistical inference. The algorithm is sufficiently complex that it would be difficult to mathematically derive these properties in a way such they could be computed from the data in a straightforward manner. It would be possible to graft these on using bootstrap methods as introduced on Section 3.6. But the bootstrap involves recomputing the estimate hundreds of times, so this may not be convenient for a relatively expensive estimator such as RANSAC. Alternatively, one could view the method as technique for identifying outliers and estimating parameters where no inference is required.

Least trimmed squares (LTS) is a method which bears some comparison to RANSAC. LTS minimizes the sum of squares of the q smallest residuals, $\sum_{i=1}^{q} \hat{\epsilon}_{(i)}^2$ where q is some number less than n and (i) indicates sorting. This method has a high breakdown point because it can tolerate a large number of outliers depending on how q is chosen. Unlike RANSAC choosing a threshold, LTS specifies the number of contaminating points we are prepared to defend against. LTS also uses a least squares criterion on the bulk of the data which has smoother properties than RANSAC.

Summary

1. Robust estimators provide protection against long-tailed errors, but they cannot overcome problems with the choice of model and its variance structure.

2. Robust estimates supply $\hat{\beta}$ and possibly standard errors without the associated inferential methods. Software and methodology for this inference require extra work.

3. Robust methods can be used in addition to least squares as a confirmatory method. You have cause to worry if the two estimates are far apart. The source of the difference should be investigated.

4. Robust estimates are useful when data need to be fit automatically without the intervention of a skilled analyst.

Exercises

1. Researchers at the National Institute of Standards and Technology (NIST) collected `pipeline` data on ultrasonic measurements of the depth of defects in the Alaska pipeline in the field. The depth of the defects were then remeasured in the laboratory. These measurements were performed in six different batches. It turns out that this batch effect is not significant and so can be ignored in the analysis that follows. The laboratory measurements are more accurate than the in-field measurements, but more time consuming and expensive. We want to develop a regression equation for correcting the in-field measurements.

   (a) Fit a regression model `Lab` ~ `Field`. Check for non-constant variance.

   (b) Use the residuals to determine an appropriate choice for the weights. Use these weights to re-estimate the model and compare with the unweighted fit.

(c) An alternative to weighting is transformation. Find transformations on `Lab` and/or `Field` so that in the transformed scale the relationship is approximately linear with constant variance. You may restrict your choice of transformation to square root, log and inverse.

2. Using the `divusa` data, fit a regression model with `divorce` as the response and `unemployed, femlab, marriage, birth` and `military` as predictors.

   (a) Make two graphical checks for correlated errors. What do you conclude?

   (b) Allow for serial correlation with an AR(1) model for the errors. What is the estimated correlation? Does the GLS model change which variables are found to be significant?

   (c) Speculate why there might be correlation in the errors.

3. For the `salmonella` dataset, fit a linear model with `colonies` as the response and `log(dose+1)` as the predictor. Check for lack of fit.

4. For the `cars` dataset, fit a linear model with `dist` as the response and `speed` as the predictor. Check for lack of fit.

5. Using the `stackloss` data, fit a model with `stack.loss` as the response and the other three variables as predictors. You can get the stackloss data using `{stackloss = sm.datasets.stackloss.load_pandas().data}`. For each of the following estimators, determine which cases have an absolute residual in excess of five.

   (a) Least squares

   (b) Least absolute deviations

   (c) Huber method

   (d) Least trimmed squares

   Compare the results — how do the coefficients vary and which estimator do you think is best in this situation?

6. Using the `cheddar` data, fit a linear model with `taste` as the response and the other three variables as predictors.

   (a) Suppose that the observations were taken in time order. Create a time variable. Plot the residuals of the model against time, and comment on what can be seen.

   (b) Fit a GLS model with the same form as above but do now allow for an AR(1) correlation among the errors. Is there evidence of such a correlation?

   (c) Fit a LS model but with time now as an additional predictor. Investigate the significance of time in the model.

   (d) The last two models have both allowed for an effect of time. Explain how they do this differently.

   (e) Suppose you were told, contrary to prior information, that the observations are not in time order. How would this change your interpretation of the model from (c)?

7. The `crawl` dataset contains data on a study looking at the age when babies learn to crawl as a function of ambient temperatures. There is additional information

about the number of babies studied each month and the variation in the response. Make an appropriate choice of weights to investigate the relationship between crawling age and temperature.

8. The `gammaray` dataset shows the x-ray decay light curve of a gamma ray burst. Build a model to predict the flux as a function time that uses appropriate weights.

9. Use the `fat` data, fitting the model described in Section 4.2.

   (a) Fit the same model, but now using Huber's robust method. Comment on any substantial differences between this model and the least squares fit.

   (b) Identify which two cases have the lowest weights in the Huber fit. What is unusual about these two points?

   (c) Plot weight (of the man) against height. Identify the two outlying cases. Are these the same as those identified in the previous question? Discuss.

Chapter 9

Transformation

Transformations of the response and/or predictors can improve the fit and correct violations of model assumptions such as nonconstant error variance. We may also consider adding additional predictors that are functions of the existing predictors like quadratic or crossproduct terms.

9.1 Transforming the Response

We start with some general considerations about transforming the response. Suppose that you are contemplating a logged response in a simple regression situation:

$$\log y = \beta_0 + \beta_1 x + \varepsilon$$

In the original scale of the response, this model becomes:

$$y = \exp(\beta_0 + \beta_1 x) \cdot \exp(\varepsilon) \tag{9.1}$$

In this model, the errors enter *multiplicatively* and not *additively* as they usually do. So the use of standard regression methods for the logged response model requires that we believe the errors enter multiplicatively in the original scale. Alternatively, for small ε, $\exp(\varepsilon) \approx 1 + \varepsilon$. Substituting this into (9.1) would result in a model with additive errors but with nonconstant variance.

If we propose the model:

$$y = \exp(\beta_0 + \beta_1 x) + \varepsilon$$

then we cannot linearize this model directly, and nonlinear regression methods might need to be applied. However, we might also use the $\exp(\varepsilon) \approx 1 + \varepsilon$ approximation to get a weighted linear model.

In practice, we may not know how the errors enter the model additively, multiplicatively or otherwise. The best approach is to try different transforms to get the structural form of the model right and worry about the error component later. We can then check the residuals to see whether they satisfy the conditions required for linear regression. If there is a problem, we have several potential solutions as discussed in earlier chapters.

Although you may transform the response, you will probably need to express predictions in the original scale. This requires back-transforming. For example, in the logged model above, your prediction would be $\exp(\hat{y}_0)$. If your prediction confidence interval in the logged scale was $[l, u]$, then you would use $[\exp l, \exp u]$. This interval will not be symmetric, but this may be desirable. For example, the untransformed prediction intervals for the Galápagos data went below zero in Section 4.1. Transformation of the response avoids this problem.

Regression coefficients will need to be interpreted with respect to the transformed scale. There is no straightforward way of back-transforming them to values that can be interpreted in the original scale. You cannot directly compare regression coefficients for models where the response transformation is different. Difficulties of this type may dissuade you from transforming the response even if this requires the use of another type of model, such as a generalized linear model.

When you use a log transformation on the response, the regression coefficients have a particular interpretation:

$$\log \hat{y} = \hat{\beta}_0 + \hat{\beta}_1 x_1 + \cdots + \hat{\beta}_p x_p$$
$$\hat{y} = e^{\hat{\beta}_0} e^{\hat{\beta}_1 x_1} \ldots e^{\hat{\beta}_p x_p}$$

An increase of one in x_1 would multiply the predicted response (in the original scale) by $e^{\hat{\beta}_1}$. Thus when a log scale is used, the regression coefficients can be interpreted in a multiplicative rather than an additive manner. A useful approximation for quick interpretations is that $\log(1 + x) \approx x$ for small values of x. So for example, suppose $\hat{\beta}_1 = 0.09$; then an increase of one in x_1 would lead to about a 0.09 increase in $\log y$ which is a 9% increase in y. The approximation is quite good up to about ± 0.25.

Box–Cox transformation

The Box–Cox method is a popular way to determine a transformation on the response. It is designed for strictly positive responses and chooses the transformation to find the best fit to the data. The method transforms the response $y \rightarrow g_\lambda(y)$ where the family of transformations indexed by λ is:

$$g_\lambda(y) = \begin{cases} \frac{y^\lambda - 1}{\lambda} & \lambda \neq 0 \\ \log y & \lambda = 0 \end{cases}$$

For fixed $y > 0$, $g_\lambda(y)$ is continuous in λ. Choose λ using maximum likelihood. The profile log-likelihood assuming normality of the errors is:

$$L(\lambda) = -\frac{n}{2} \log(\text{RSS}_\lambda / n) + (\lambda - 1) \sum \log y_i$$

where RSS_λ is the residual sum of squares when $g_\lambda(y)$ is the response. You can compute $\hat{\lambda}$ numerically to maximize this. If the purpose of the regression model is prediction, then use y^λ as the response (no need to use $(y^\lambda - 1)/\lambda$, as the rescaling is just so $g_\lambda \rightarrow \log$ as $\lambda \rightarrow 0$ which maintains continuity in the likelihood). If explaining the model is important, you should round λ to the nearest interpretable value. For example, if $\hat{\lambda} = 0.46$, it would be hard to explain what this new response means, but $\sqrt{y}$ might be easier.

Transforming the response can make the model harder to interpret, so we do not want to do it unless it is really necessary. One way to check this is to form a confidence interval for λ. A $100(1 - \alpha)\%$ confidence interval for λ is:

$$\{\lambda: \quad L(\lambda) > L(\hat{\lambda}) - \frac{1}{2}\chi_1^{2(1-\alpha)}\}$$

This interval can be derived by inverting the likelihood ratio test of the hypothesis that $H_0 : \lambda = \lambda_0$ which uses the statistic $2(L(\hat{\lambda}) - L(\lambda_0))$ having approximate null distribution χ_1^2. The confidence interval also tells you how much it is reasonable to round λ for the sake of interpretability.

We check whether the response in the savings data needs transformation. Start by loading the packages:

```
import pandas as pd
import numpy as np
import matplotlib.pyplot as plt
import scipy as sp
import statsmodels.api as sm
import statsmodels.formula.api as smf
import faraway.utils
```

Try it out on the savings dataset and plot the results:

```
import faraway.datasets.savings
savings = faraway.datasets.savings.load()
lmod = smf.ols('sr ~ pop15 + pop75 + dpi + ddpi',
    savings).fit()
X = lmod.model.wexog
n = savings.shape[0]
sumlogy = np.sum(np.log(savings.sr))
lam = np.linspace(0.5,1.5,100)
llk = np.empty(100)
for i in range(0, 100):
    lmod = sm.OLS(sp.stats.boxcox(savings.sr,lam[i]),X).fit()
    llk[i] = -(n/2)*np.log(lmod.ssr/n) + (lam[i]-1)*sumlogy
fig, ax = plt.subplots()
ax.plot(lam,llk)
ax.set_xlabel('$\lambda$')
ax.set_ylabel('log likelihood')
maxi = llk.argmax()
ax.vlines(lam[maxi],ymin=min(llk),ymax=max(llk),
    linestyle = 'dashed')
cicut = max(llk) - sp.stats.chi2.ppf(0.95,1)/2
rlam = lam[llk > cicut]
ax.hlines(cicut,xmin=rlam[0],xmax=rlam[-1],linestyle = 'dashed')
ax.vlines([rlam[0],rlam[-1]],ymin=min(llk),ymax=cicut,
    linestyle = 'dashed')
```

The plot is shown in Figure 9.1. In the computation above, we chose the search range for λ as [0.5,1.5]. You may find you need to make it wider than this to capture the maximum and the confidence interval. We admit to have chosen this range on the basis of some prior experimentation. The confidence interval for λ runs from about 0.6 to about 1.4. We can see that there is no good reason to transform.

Now consider the Galápagos Islands dataset analyzed earlier:

```
import faraway.datasets.galapagos
galapagos = faraway.datasets.galapagos.load()
from patsy import dmatrix
```

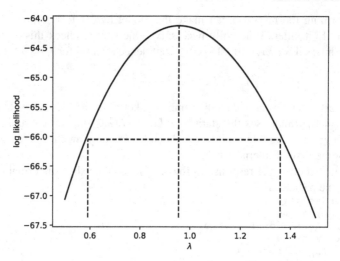

Figure 9.1   Log-likelihood plot for the Box–Cox transformation of the savings data.

```python
X = dmatrix('Area + Elevation + Nearest + Scruz + Adjacent',
    data=galapagos)
n = galapagos.shape[0]
sumlogy = np.sum(np.log(galapagos.Species))
lam = np.linspace(-0.0,0.65,100)
llk = np.empty(100)
for i in range(0, 100):
    lmod = sm.OLS(sp.stats.boxcox(galapagos.Species,
                  lam[i]),X).fit()
    llk[i] = -(n/2)*np.log(lmod.ssr/n) + (lam[i]-1)*sumlogy
fig, ax = plt.subplots()
ax.plot(lam,llk)
ax.set_xlabel('$\lambda$')
ax.set_ylabel('log likelihood')
maxi = llk.argmax()
ax.vlines(lam[maxi],ymin=min(llk),ymax=max(llk),
    linestyle = 'dashed')
cicut = max(llk) - sp.stats.chi2.ppf(0.95,1)/2
rlam = lam[llk > cicut]
ax.hlines(cicut,xmin=rlam[0],xmax=rlam[-1],linestyle = 'dashed')
ax.vlines([rlam[0],rlam[-1]],ymin=min(llk),ymax=cicut,
    linestyle = 'dashed')
```

The plot is shown in the left panel of Figure 9.2. Again we have selected the interval for λ using some experimentation to catch a good window around the maximum. We see that perhaps a cube root transformation might be best here. A square root is also a possibility, as this falls just within the confidence intervals. Certainly there is a strong need to transform.

Some general considerations concerning the Box–Cox method are:

1. The Box–Cox method gets upset by outliers — if you find $\hat{\lambda} = 5$, then this is

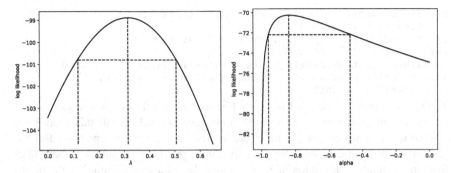

Figure 9.2   Log-likelihood plot for the Box–Cox transformation of the Galápagos data is shown on the left and the log-additive transformation on the leafburn data is shown on the right.

probably the reason — there can be little justification for actually making such an extreme transformation.

2. If some $y_i < 0$, we can add a constant to all the y. This can work provided the constant is small, but this is an inelegant solution.

3. If $\max_i y_i / \min_i y_i$ is small, then the Box–Cox will not have much real effect because power transforms are well approximated by linear transformations over short intervals far from the origin.

4. There is some doubt whether the estimation of λ counts as an extra parameter to be considered in the degrees of freedom. This is a difficult question since λ is not a linear parameter and its estimation is not part of the least squares fit.

The Box–Cox method is not the only way of transforming the predictors. Another family of transformations is given by $g_\alpha(y) = \log(y + \alpha)$. We can illustrate the value of this using some data from Steel and Torrie (1980) on the burn time of tobacco leaves as a function of three chemical constituents: nitrogen, chlorine and potassium. The method of calculation follows the same path:

```
import faraway.datasets.leafburn
leafburn = faraway.datasets.leafburn.load()
from patsy import dmatrix
X = dmatrix('nitrogen + chlorine + potassium',data=leafburn)
n = leafburn.shape[0]
alpha = np.linspace(-0.999,0,100)
llk = np.empty(100)
for i in range(0, 100):
    lmod = sm.OLS(np.log(leafburn.burntime+alpha[i]),X).fit()
    llk[i] = -(n/2)*np.log(lmod.ssr) - \
        np.sum(np.log(leafburn.burntime + alpha[i]))
fig, ax = plt.subplots()
ax.plot(alpha,llk)
ax.set_xlabel(r'$\alpha$')
ax.set_ylabel('log likelihood')
maxi = llk.argmax()
```

```
ax.vlines(alpha[maxi],ymin=min(llk),ymax=max(llk),
    linestyle = 'dashed')
cicut = max(llk) - sp.stats.chi2.ppf(0.95,1)/2
ralp = alpha[llk > cicut]
ax.hlines(cicut,xmin=ralp[0],xmax=ralp[-1],linestyle = 'dashed')
ax.vlines([ralp[0],ralp[-1]],ymin=min(llk),ymax=cicut,
    linestyle = 'dashed')
```

The plot is shown in the right panel of Figure 9.2. The Box-Cox analysis of this model recommends $\hat{\lambda} = -0.34$ and the confidence interval bounds this away from zero so it seems we cannot use the more interpretable log transformation. But the plot shows that we can use a log transformation provided we subtract about 0.85 from the response. This value may be interpretable as a start-up time for the fire to get going.

For responses that are proportions (or percentages), the logit transformation, $\log(y/(1-y))$, is often used, while for responses that are correlations, Fisher's z transform, $y = 0.5\log((1+y)/(1-y))$, is worth considering.

9.2   Transforming the Predictors

You can take a Box–Cox style approach for each of the predictors, choosing the transformation to minimize the RSS. However, this usually is not the best way to find good transformations on the predictors. You can also use graphical methods such as partial residual plots to select transforming the predictors. These methods are designed to replace x in the model with f(x) for some chosen f. The methods we consider below are more general in that they replace x with more than one term — f(x) + g(x) + ... This allows more flexibility since each additional term carries a parameter.

9.3   Broken Stick Regression

Sometimes we have reason to believe that different linear regression models apply in different regions of the data. For example, in the analysis of the savings data, we observed that there were two groups in the data and we might want to fit a different model to the two parts. Suppose we focus attention on just the pop15 predictor for ease of presentation. We fit the two regression models depending on whether pop15 is greater or less than 35%. The two fits are seen in Figure 9.3.

```
lmod1 = smf.ols('sr ~ pop15', savings[savings.pop15 < 35]).fit()
lmod2 = smf.ols('sr ~ pop15', savings[savings.pop15 > 35]).fit()
plt.scatter(savings.pop15, savings.sr)
plt.xlabel('Population under 15')
plt.ylabel('Savings rate')
plt.axvline(35,linestyle='dashed')
plt.plot([20,35],[lmod1.params[0]+lmod1.params[1]*20,
    lmod1.params[0]+lmod1.params[1]*35],'k-')
plt.plot([35,48],[lmod2.params[0]+lmod2.params[1]*35,
    lmod2.params[0]+lmod2.params[1]*48],'k-')
```

A possible objection to this subsetted regression fit is that the two parts of the fit do not meet at the join. If we believe the fit should be continuous as the predictor

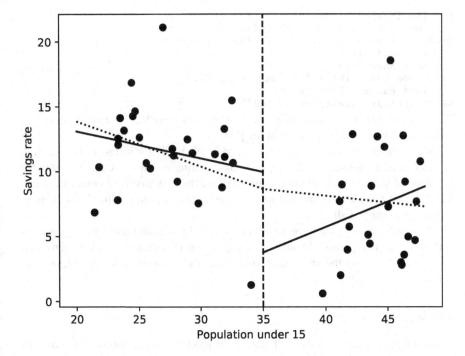

Figure 9.3   Subset regression fit is shown with the solid line, while the broken stick regression is shown with the dotted line.

varies, we should consider the broken stick regression fit. Define two *basis functions*:

$$B_l(x) = \begin{cases} c - x & \text{if } x < c \\ 0 & \text{otherwise} \end{cases}$$

and:

$$B_r(x) = \begin{cases} x - c & \text{if } x > c \\ 0 & \text{otherwise} \end{cases}$$

where c marks the division between the two groups. B_l and B_r form a first-order spline basis with a knotpoint at c. Sometimes B_l and B_r are called hockey-stick functions because of their shape. We can now fit a model of the form:

$$y = \beta_0 + \beta_1 B_l(x) + \beta_2 B_r(x) + \varepsilon$$

using standard regression methods. The two linear parts are guaranteed to meet at c. Notice that this model uses only three parameters in contrast to the four total parameters used in the subsetted regression illustrated before. A parameter has been saved by insisting on the continuity of the fit at c.

We define the two hockey-stick functions, compute and display the fit:

```
def lhs (x,c): return(np.where(x < c, c-x, 0))
def rhs (x,c): return(np.where(x < c, 0, x-c))
lmod = smf.ols('sr ~ lhs(pop15,35) + rhs(pop15,35)',
    savings).fit()
x = np.arange(20,49)
py = lmod.params[0] + lmod.params[1]*lhs(x,35) + \
    lmod.params[2]*rhs(x,35)
plt.plot(x,py,linestyle='dotted')
```

The two (dotted) lines now meet at 35, as shown in Figure 9.3. The intercept of this model is the value of the response at the join.

We might question which fit is preferable in this particular instance. For the high pop15 countries, we see that the imposition of continuity causes a change in sign for the slope of the fit. We might argue that because the two groups of countries are so different and there are so few countries in the middle region, we might not want to impose continuity at all.

We can have more than one knotpoint simply by defining more basis functions with different knotpoints. Broken stick regression is sometimes called *segmented regression*. Allowing the knotpoints to be parameters is worth considering, but this will result in a nonlinear model.

9.4 Polynomials

Another way of generalizing the $X\beta$ part of the model is to add polynomial terms. In the one-predictor case, we have:

$$y = \beta_0 + \beta_1 x + \cdots + \beta_d x^d + \varepsilon$$

which allows for a more flexible relationship. We usually do not believe the polynomial exactly represents any underlying reality, but it can allow us to model expected features of the relationship. A quadratic term allows for a predictor to have an optimal setting. For example, there may be a best temperature for baking bread – a hotter or colder temperature may result in a less tasty outcome. If you believe a predictor behaves in this manner, it makes sense to add a quadratic term.

In the following example, ethanol fuel was burned in a single-cylinder engine. The emission of NOx varied according to the engine compression(C) and equivalence ratio(E). We can obtain the data with:

```
ethanol = sm.datasets.get_rdataset("ethanol", "lattice").data
```

The relationship between C and NOx as seen in Figure 9.4 has an obvious maximum, so just a linear term for this predictor will not suffice. The simplest model uses a quadratic term for E:

```
lmod2 = smf.ols('NOx ~ E + I(E**2) + C', ethanol).fit()
lmod2.sumary()
```

	coefs	stderr	tvalues	pvalues
Intercept	-21.203	1.240	-17.10	0.0000
E	52.411	2.704	19.39	0.0000
I(E ** 2)	-29.090	1.478	-19.68	0.0000
C	0.064	0.014	4.63	0.0000

n=88 p=4 Residual SD=0.484 R-squared=0.82

We might want to know the value of E that leads to maximum output of NOx. We can use simple calculus to maximize the quadratic:

```
-lmod2.params[1]/(2*lmod2.params[2])
0.9008
```

Although it's not necessary here, we can use numerical optimization to find the maximum:

```
f2 = np.poly1d(lmod2.params[[2,1,0]])
from scipy.optimize import minimize_scalar
result = minimize_scalar(-f2)
result.x
0.9008
```

We define the polynomial using the `poly1d()` function noting that the coefficients are expected with the highest degree first. Since the {`minimize_scalar()`} function computes the minimum, we need to negate the function to get the maximum. Reassuringly, the result is the same. Notice that we are able to perform this optimization without reference to C since the model uses the predictors in a separable way.

We might check that the model fits the data well with a residual check such as:

```
plt.scatter(ethanol.E, lmod2.resid)
```

The plot (not shown) indicates some remaining structure that suggests a more flexible function than the quadratic is required. We might prefer a quartic rather than a cubic to get the shape we expect for predicting NOx:

```
lmod4 = smf.ols('NOx ~ E + I(E**2) +I(E**3) + I(E**4) + C',
        ethanol).fit()
lmod4.sumary()
```

	coefs	stderr	tvalues	pvalues
Intercept	161.782	16.249	9.96	0.0000
E	-821.333	77.013	-10.66	0.0000
I(E ** 2)	1,496.858	133.971	11.17	0.0000
I(E ** 3)	-1,156.081	101.392	-11.40	0.0000
I(E ** 4)	321.154	28.205	11.39	0.0000
C	0.055	0.009	6.27	0.0000

```
n=88 p=6 Residual SD=0.305 R-squared=0.93
```

The residual checks are much more satisfactory and the fit, judging by the R-squared, is certainly better. Yet there are some drawbacks. The estimated coefficients are very large. Sometimes this happens because taking higher powers of numbers tends to make them either very large (if much greater than one in absolute value) or very small (if much smaller than one). The corresponding coefficients will need to compensate for this to achieve a sensible response. But this is not the problem with this data. The large values of the coefficients are due to the strong collinearities between the polynomial terms. This has negative consequences for the stability of the estimation and predictions.

In this example, we observe that the predicted response cannot be negative. A simple way to ensure this is to use a logged response:

```
lmodl = smf.ols('np.log(NOx) ~ E + I(E**2) + C', ethanol).fit()
lmodl.sumary()
```

	coefs	stderr	tvalues	pvalues
Intercept	-13.935	0.519	-26.87	0.0000
E	32.799	1.131	29.01	0.0000

```
I(E ** 2) -18.165  0.618   -29.38  0.0000
C           0.030  0.006     5.17  0.0000
```

```
n=88 p=4 Residual SD=0.202 R-squared=0.91
```

This fit is superior to the unlogged quadratic model.

Let's define the functions associated with the quartic and logged quadratic model:
```
f4 = np.poly1d(lmod4.params[[4,3,2,1,0]])
fl = np.poly1d(lmodl.params[[2,1,0]])
```

Now we can plot all three fits with:
```
plt.scatter(ethanol.E,ethanol.NOx,label=None)
z = np.linspace(0.4,1.4)
meanC = np.mean(ethanol.C)
plt.plot(z,f2(z)+meanC*lmod2.params['C'],'k-',label='quadratic')
plt.plot(z,f4(z)+meanC*lmod4.params['C'],'k--',label='quartic')
plt.plot(z,np.exp(fl(z)+meanC*lmodl.params['C']),'k:',
    label='log-quadratic')
plt.legend()
```

I have chosen to compute the prediction on a range of E values slightly wider than that observed in the data. I use the mean value of C. We can see the resulting fit in Figure 9.4. We see that all three models fit the data quite well but their behavior

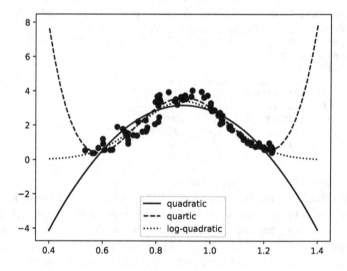

Figure 9.4   Polynomial fits to the ethanol engine data. All three fitted lines use the mean value of C.

is unstable as we extrapolate. Polynomial functions are dominated by the term of highest order as the argument becomes more extreme. They will inevitably produce unrealistic extrapolations. For higher order polynomials, the problem will be worse and occur for even modest extrapolations. In this example, we are able to exploit the logged response to ensure sensible behavior away from the bulk of the data, but this will not always be feasible.

Now let's attempt to find the optimizing value of E for the quartic model:
```
minimize_scalar(-f4).x
```
This results in a surge of warnings and an implausibly large value. From Figure 9.4, we see we must restrict the range of the maximization to capture the central maximum:
```
minimize_scalar(-f4,bounds=(0.6,1.2),method="bounded").x
0.9017
```
For completeness, here is the maximum under the logged quadratic model:
```
minimize_scalar(-f1).x
0.9028
```
Although the three models are apparently quite different, the maximizing value is very similar. Given the symmetry seen in the data, this is not surprising.

The message from this example is that using polynomials of order higher than a quadratic is problematic. The problems with collinearity and unstable extrapolation will only get worse as the degree is increased. The estimated coefficients have no useful interpretation. If you want a more flexible fit, I advise you to use other methods such as the piecewise linear fit of the broken-stick regression or the spline methods described below.

If you really insist on using polynomials, you should consider the use of Legendre polynomials. These are polynomials $P_n(x)$ of degree n defined on $[-1, 1]$ which have the orthogonal property:

$$\int_{-1}^{1} P_m(x)P_n(x)dx = 0, \qquad \forall m \neq n$$

If we normalize by requiring that $\int_{-1}^{1} P_n(x) = 1$, this is sufficient to uniquely define them. The first few ones are $1, x, (3x^2 - 1)/2, (5x^3 - 3x)/2$.

Here we use Legendre polynomials rather than a standard polynomial up the fourth degree:
```
cmax = max(ethanol.E)
cmin = min(ethanol.E)
a=2/(cmax-cmin)
b=-(cmax+cmin)/(cmax-cmin)
sC = ethanol.E * a + b
X = np.polynomial.legendre.legvander(sC,4)
Ccol= np.array(ethanol['C']).reshape(-1,1)
X = np.concatenate((X,Ccol),axis=1)
lmodleg = sm.OLS(ethanol.NOx,X).fit()
lmodleg.sumary()
```

	coefs	stderr	tvalues	pvalues
const	1.330	0.111	12.03	0.0000
x1	0.271	0.059	4.59	0.0000
x2	-2.413	0.079	-30.39	0.0000
x3	-0.358	0.086	-4.15	0.0001
x4	1.083	0.095	11.39	0.0000
x5	0.055	0.009	6.27	0.0000

```
n=88 p=6 Residual SD=0.305 R-squared=0.93
```
We first need to scale the predictor E to the range $[-1, 1]$. We create the design matrix using the legvander function and remember to append the column for C. The fitted values are the same as the quartic model above, but the coefficients are now stable.

We can compute the correlations between the Legendre polynomial terms with:
```
np.corrcoef(X[:,1:5].T).round(3)
array([[1.   , 0.303, 0.076, 0.17 ],
       [0.303, 1.   , 0.29 , 0.232],
       [0.076, 0.29 , 1.   , 0.35 ],
       [0.17 , 0.232, 0.35 , 1.   ]])
```
We see that these correlations are generally small. If the predictor values were evenly spaced, the correlations would be practically zero due to the defining property of Legendre polynomials. It would be desirable to have zero correlation since it would result in *orthogonality*. This is particularly useful if one wants to choose the appropriate degree for the polynomials. With an orthogonal design, the coefficients do not change as other terms are added or removed from the model. This makes it possible to choose the degree with a single fit of the model.

It is possible to create an orthogonal design even with an unevenly spaced predictor, but one must question whether this is worth the effort. Orthogonal polynomials were popular years ago when computing was expensive. Using such a model also requires additional effort with keeping track of scaling. The benefits do not outweigh the costs.

Usually it is a bad idea to eliminate lower order terms from the model before the higher order terms even if they are not statistically significant. This follows from the *hierarchy principle*. Usually, the removal of an unneeded input from a model is beneficial since less information is required to make predictions. But in this case, we still need the input so no such advantage is gained. Futhermore, an additive change in scale would change the t-statistic of all but the highest order term. We would not want the conclusions of our study to be sensitive to such changes in the scale which ought to be inconsequential. For example, suppose we transform ddpi by subtracting 10 and refit the quadratic model:
```
ethanol['Ec'] = ethanol.E - 0.9
lmodc = smf.ols('NOx ~ Ec + I(Ec**2) + C', ethanol).fit()
lmodc.sumary()
            coefs stderr tvalues pvalues
Intercept   2.404  0.168   14.32  0.0000
Ec          0.049  0.257    0.19  0.8488
I(Ec ** 2) -29.090 1.478  -19.68  0.0000
C           0.064  0.014    4.63  0.0000

n=88 p=4 Residual SD=0.484 R-squared=0.82
```
We see that the quadratic term remains unchanged from the previous uncentered form of this model, but the linear term is now insignificant. There is no good reason to remove the linear term in this model, but not in the previous version. If you do remove lower order terms from, say, a quadratic model, be aware that this has some special meaning. Setting the intercept to zero means the regression passes through the origin while setting the linear term to zero would mean that the response is optimized at a predictor value of zero.

You can also define polynomials in more than one variable. These are sometimes called *response surface* models. A second degree model would be:

$$y = \beta_0 + \beta_1 x_1 + \beta_2 x_2 + \beta_{11} x_1^2 + \beta_{22} x_2^2 + \beta_{12} x_1 x_2$$

The PolynomialFeatures() function from scikit-learn is useful for constructing the polynomial basis. We choose a second-order model. Note the order of the terms which we have labeled.

```
from sklearn.preprocessing import PolynomialFeatures
poly = PolynomialFeatures(2)
X = ethanol.loc[:,['C','E']]
Xp = poly.fit_transform(X)
Xp = pd.DataFrame(Xp,columns=['Intercept','C','E','C2','E2','CE'])
lmod = sm.OLS(ethanol.NOx,Xp).fit()
lmod.sumary()
```

```
          coefs stderr tvalues pvalues
Intercept -24.262  1.545  -15.70  0.0000
C           0.224  0.119    1.88  0.0642
E          56.757  2.756   20.59  0.0000
C2          0.002  0.004    0.56  0.5737
E2         -0.239  0.062   -3.87  0.0002
CE        -29.785  1.388  -21.46  0.0000

n=88 p=6 Residual SD=0.450 R-squared=0.85
```

All the terms are strongly significant except for the quadratic term of C. This is apparent in the fitted surface seen in Figure 9.5 where very little curvature is seen in the C direction. The CE term is often called the interaction term and represents the rotation of the surface relative to the axes. Without this term, we might freely optimize one variable without concern for the other as we did in the previous analysis. This model shows that we cannot reasonably maximize E without reference to C.

Although the quadratic term in C is not significant, we would gain little from removing it from the model. We respect the *hierarchy* principle since other terms of second order E2 and CE appear significantly in the model. This model also shows the simplest way of allowing for an interaction between two quantitative predictors. It would not make much sense to have just CE in the model. Including the interaction means including all three second-order terms.

We can construct a perspective wireframe plot of the fitted surface as follows. We compute the fit on a 11×11 grid of values covering the range of the predictors and see the result in Figure 9.5.

```
ngrid = 11
Cx = np.linspace(min(ethanol.C),max(ethanol.C),ngrid)
Ey = np.linspace(min(ethanol.E),max(ethanol.E),ngrid)
X, Y = np.meshgrid(Cx,Ey)
XYpairs = np.dstack([X, Y]).reshape(-1, 2)
Xp = poly.fit_transform(XYpairs)
pv = np.dot(Xp,lmod.params)
Z = np.reshape(pv,(ngrid,ngrid))
from mpl_toolkits.mplot3d import Axes3D
fig = plt.figure()
ax = fig.add_subplot(111, projection='3d')
ax.plot_wireframe(X,Y,Z)
ax.set_xlabel("C")
ax.set_ylabel("E")
ax.set_zlabel("NOx")
```

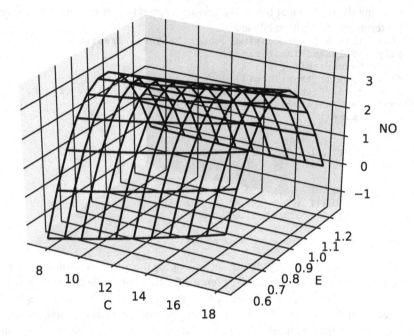

Figure 9.5   Perspective plot of the quadratic surface fit to the saving data.

9.5   Splines

Polynomials have the advantage of smoothness, but the disadvantage is that each data point affects the fit globally. This is because the power functions used for the polynomials take nonzero values across the whole range of the predictor. In contrast, the broken stick regression method localizes the influence of each data point to its particular segment which is good, but we do not have the same smoothness as with the polynomials. There is a way we can combine the beneficial aspects of both these methods — smoothness and local influence — by using *B-spline* basis functions.

We can see why splines might be helpful by using a simulated example. Suppose we know the true model is:

$$y = \sin^3(2\pi x^3) + \varepsilon, \qquad \varepsilon \sim N(0, (0.1)^2)$$

The advantage of using simulated data is that we can see how close our methods come to the truth. We generate the data and display them in the first plot of Figure 9.6.

```
def funeg (x): return(np.sin(2*np.pi*x**3)**3)
x = np.linspace(0., 1., 101)
y = funeg(x) + sp.stats.norm.rvs(0,0.1,101)
plt.scatter(x,y)
plt.plot(x, funeg(x))
```

We see how polynomial fits of degrees 4 and 12 do in fitting these data:

```
p = np.poly1d(np.polyfit(x,y,4))
```

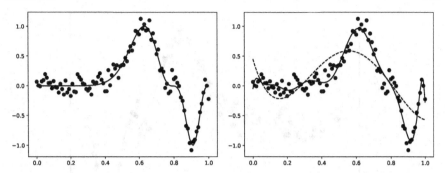

Figure 9.6 Data and true function shown on the left. Orthogonal polynomial of order 4 (dashed) and order 12 (solid) shown on the right.

```
plt.scatter(x,y)
plt.plot(x,p(x),'k--')
p = np.poly1d(np.polyfit(x,y,12))
plt.plot(x,p(x),'k-')
```

The two fits are shown in the second panel of Figure 9.6. We see that degree 4 is a clear underfit; order 12 is much better, although the fit is too wiggly in the first section and misses the point of inflection around $x = 0.8$.

We may define a cubic B-spline basis on the interval $[a,b]$ by the following requirements on the interior basis functions with knotpoints at $t_1, \dots, t_k$:

1. A given basis function is nonzero on an interval defined by four successive knots and zero elsewhere. This property ensures the local influence property.

2. The basis function is a cubic polynomial for each subinterval between successive knots.

3. The basis function is continuous and is also continuous in its first and second derivatives at each knotpoint. This property ensures the smoothness of the fit.

4. The basis function integrates to one over its support.

The basis functions at the ends of the interval are defined a little differently to ensure continuity in derivatives at the edge of the interval. A full definition of B-splines and more details about their properties may be found in de Boor (2002). The broken stick regression is an example of the use of linear splines.

Regression splines are useful for fitting functions with some flexibility provided we have enough data. We can form basis functions for all the predictors in our model, but we need to be careful not to use up too many degrees of freedom.

```
from patsy import bs
kts = [0,0.2,0.4,0.5,0.6,0.7,0.8,0.85,0.9,1]
z = sm.OLS(y,bs(x,knots=kts,include_intercept = True)).fit()
plt.scatter(x,y)
plt.plot(x, z.fittedvalues)
```

A related alternative to regression splines is smoothing splines. Suppose the model is $y_i = f(x_i) + \varepsilon_i$, so in the spirit of least squares, we might choose $\hat{f}$ to

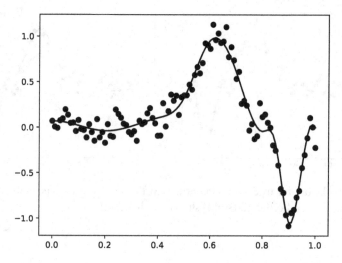

Figure 9.7   Regression spline fit to the simulated data.

minimize the sum of squares. Unfortunately, the unhelpful solution is $\hat{f}(x_i) = y_i$. This "join the dots" regression is almost certainly too rough. Instead, suppose we choose $\hat{f}$ to minimize a modified least squares criterion:

$$\frac{1}{n}\sum(Y_i - f(x_i))^2 + \lambda \int [f''(x)]^2 dx$$

where $\lambda > 0$ controls the amount of smoothing and $\int [f''(x)]^2 dx$ is a *roughness penalty*. When f is rough, the penalty is large, but when f is smooth, the penalty is small. Thus the two parts of the criterion balance fit against smoothness. This is the *smoothing spline* fit.

While recovery of an unknown function from data is an intriguing mathematical challenge, it is important not to get diverted from the goal of the analysis. Where explanation is the purpose, we may be looking for some supposed feature of the function such as a maximum or point of inflexion. For such an aim, one needs to focus rather than simply to attempt to estimate the function in general.

9.6   Additive Models

Searching for good transformations on the predictors is difficult when there are multiple predictors. Changing the transformation on one predictor may change the best choice of transformation on another predictor. Fortunately, there is a way to simultaneously choose the transformations. An *additive* model takes the form :

$$y = \alpha + f_1(X_1) + f_2(X_2) + \cdots + f_p(X_p) + \varepsilon$$

The linear terms of the form $\beta_i X_i$ have been replaced with more flexible functional forms $f_i(X_i)$. Models that can handle a nonnormal response are called *Generalized*

Additive Models (GAM). The `statsmodels` package has some GAM fitting capability although we do need the generalized component here. We demonstrate the methodology on the ethanol data analyzed earlier. We need to specify the general functional forms for f_1 and f_2. B-splines are a common choice with cubic (degree=3) being standard. We also need to decide how many splines to use in each dimension. For a relatively small dataset such as this, a smaller number, such as 4, is appropriate. Larger datasets would justify the greater flexibility accorded by using more splines. The fitting process requires that we set up the B-spline basis before making the fit:

```
from statsmodels.gam.api import GLMGam, BSplines
xmat = ethanol[['C', 'E']]
bs = BSplines(xmat, df=[4, 4], degree=[3, 3])
gamod = GLMGam.from_formula('NOx ~ C + E', ethanol,
    smoother=bs).fit()
```

We could examine the fitted coefficients from this model, but these are not easily interpreted. Instead it is more valuable to plot the predicted response as a function of one variable with other variables held fixed as we see in the pair of plots in Figure 9.8. Partial residuals and 95% confidence bands are also displayed.

```
fig=gamod.plot_partial(0, cpr=True)
fig=gamod.plot_partial(1, cpr=True)
```

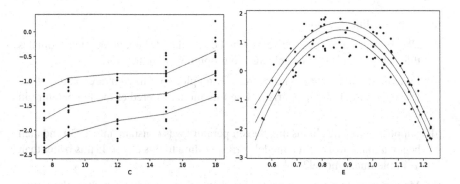

Figure 9.8   Fitted functions using the additive model for the ethanol data. The 95% confidence bands are also shown.

We see a roughly linear relationship for C and a quadratic one for E. This confirms the conclusions of the previous analyses. Of course, the GAM analysis is more valuable when we do not know what to expect in advance, especially when there are more than two predictors.

The GAM approach can be used as a complete system of inference in its own right or as an exploratory technique to discover good transformations on the predictors for linear models. The current state of the software in Python means that only the latter approach is currently viable.

Some care is necessary in choosing the amount of smoothing to use in a GAM. In our example, the default choice is satisfactory. The GAM functions within `statsmodels` have the ability to choose the amount of smoothing according to various criteria.

9.7 More Complex Models

Regression analysis relies on the skill of the human analyst to make judgments about graphical displays and to incorporate subject area knowledge. When the purpose of the analysis is explanation or the sample size is relatively small, regression analysis compares well to more complex alternatives. The linear model produces interpretable parameters which are essential if we want to gain some understanding of the relationship between the variables. If we do not have much data, it is hard to justify a more complex approach.

For larger datasets where prediction is the goal, more complex models using methods from machine learning may be more effective when the primary goal is prediction. This is because these methods will be able to fit the data more flexibly while keeping the number of parameters under control.

Exercises

1. The `aatemp` data come from the U.S. Historical Climatology Network. They are the annual mean temperatures (in degrees F) in Ann Arbor, Michigan going back about 150 years.

   (a) Is there a linear trend?

   (b) Observations in successive years may be correlated. Fit a model that estimates this correlation. Does this change your opinion about the trend?

   (c) Fit a polynomial model with degree 10 and plot your fitted model on top of the data. Do any difficulties arise? Use this model to predict the temperature in 2020.

   (d) Suppose someone claims that the temperature was constant until 1930 and then began a linear trend. Fit a model corresponding to this claim. Is this better than a simple linear model?

   (e) Make a cubic spline fit with six basis functions evenly spaced on the range. Plot the fit in comparison to the previous fits. Does this model fit better than the straight-line model?

2. The `cornnit` data on the relationship between corn yield (bushels per acre) and nitrogen (pounds per acre) fertilizer application were studied in Wisconsin in 1994. Use transformations to find a good model for predicting yield from nitrogen. Use a goodness-of-fit test to check your model. *Tip:* The response is called `yield` which is a keyword in Python. Use the `pandas, rename()` function to avoid unexpected problems.

3. Using the `ozone` data, fit a model with O3 as the response and `temp`, `humidity` and `ibh` as predictors. Use the Box–Cox method to determine the best transformation on the response.

4. Use the `pressure` data to fit a model with `pressure` as the response and `temperature` as the predictor using transformations to obtain a good fit using informal exploratory methods.

5. Use transformations to find a good model for `volume` in terms of `girth` and `height` using the `trees` data. Think about the formula for the volume of a cylinder and use this as the starting point. Plot the transformed data in 2D and 3D. You can load the data with:

    `trees = sm.datasets.get_rdataset("trees", "datasets")`

    with the data found as `trees.data` and documentation seen with
    `{print(trees.__doc__}.`

6. Use the `odor` data for this question.

    (a) Fit a second order response surface for the `odor` response using the other three variables as predictors. How many parameters does this model use and how many degrees of freedom are left?

    (b) Fit a model for the same response but now excluding any interaction terms but including linear and quadratic terms in all three predictors. Compare this model to the previous one. Is this simplification justified?

    (c) Use the previous model to determine the values of the predictors which result in the minimum predicted odor.

7. Use the `cheddar` data for this question.

    (a) Fit an additive model for a response of `taste` with the other three variables as predictors. Plot the residuals against the fitted values and each of the predictors. Is any transformation of the predictors suggested?

    (b) Use the Box–Cox method to determine an optimal transformation of the response. Would it be reasonable to leave the response untransformed?

    (c) Use the optimal transformation of the response and refit the additive model. Does this make any difference to the transformations suggested for the predictors?

8. Use the `cars` data with distance as the response and speed as the predictor.

    (a) Plot distance against speed.

    (b) Show a linear fit to the data on the plot.

    (c) Show a quadratic fit to the data on the plot.

    (d) Now use `sqrt(dist)` as the response and fit a linear model. Show the fit on the same plot.

Chapter 10

Model Selection

For all but the simplest cases we are confronted with a choice of possible regression models for our data. We may even have expanded the choice of possible models by introducing new variables derived from those available originally by making transformations, creating interactions or adding polynomial terms. In the machine learning world, the predictors are often called *features*. The process of creating new variables or features is called *feature engineering*. In this context, model selection might be called *feature selection*.

Model selection is not always necessary. Sometimes there is one established model derived from physical theory or empirical experience. In other cases, we commit to a particular model as part of a designed experiment. This would be usual in clinical trials. We might consider whether this model is adequate given the data but we will not start by considering alternative models. Also, if the number of predictors is relatively small compared to the number of observations, there may not be much benefit from model selection and it might be justifiable not to trouble with it.

Let's suppose we do want to actively choose a model. Our problem is then in selecting the "best" subset of predictors. More information should only be helpful so one might wonder why we do not simply include all the available variables in the model. However, we may wish to consider a smaller model for several reasons. The principle of Occam's Razor states that among several plausible explanations for a phenomenon, the simplest is best. Applied to regression analysis, this implies that the smallest model that fits the data adequately is best.

Another consideration is that unnecessary predictors will add noise to the estimation of other quantities that interested us. Degrees of freedom will be wasted. More precise estimates and predictions might be achieved with a smaller model. In some cases, collecting data on additional variables can cost time or money so a smaller prediction model may be more economical.

Model selection is a process that should not be separated from the rest of the analysis. Other parts of the data analysis can have an impact. For example, outliers and influential points can do more than just change the current model — they can change the model we select. It is important to identify such points. Also transformations of the variables can have an impact on the model selected. Some iteration and experimentation are often necessary to find better models.

Although Occam's Razor is a compelling heuristic, we must focus our effort on the main objective of regression modeling. We might obtain better predictions by using larger models so although smaller models might be appealing, we do not wish to compromise on predictive ability. For investigations that focus on the explanatory effect of the predictors, one should be cautious about the use of automated variable selection procedures. In such cases, attention is put on just a few predictors of interest, while the remaining predictors are not of primary interest but must be controlled for. It would be unwise to expect an automated procedure to do this reliably.

Even with a moderate number of potential predictors, the possible combinations of variables can become very large. Procedures that consider all possible combinations may not be practical and we must step through the space of possible models in an incremental way. In some situations, the model space is structured hierarchically which constrains the reasonable choice of model.

10.1 Hierarchical Models

Some families of models have a natural hierarchy. For example, in polynomial models, x^2 is a higher order term than x. When selecting variables, it is important to respect the hierarchy. Lower order terms should not usually be removed from the model before higher order terms in the same variable. We introduced the hierarchy principle in Section 9.4 and illustrated it with application to polynomials.

For models with interactions, consider the example of a second-order response surface model:

$$y = \beta_0 + \beta_1 x_1 + \beta_2 x_2 + \beta_{11} x_1^2 + \beta_{22} x_2^2 + \beta_{12} x_1 x_2 + \varepsilon$$

We would not normally consider removing the $x_1 x_2$ interaction term without simultaneously considering the removal of the x_1^2 and x_2^2 terms. A joint removal would correspond to the clearly meaningful comparison of a quadratic surface and a linear one. Just removing the $x_1 x_2$ term would correspond to a surface that is aligned with the coordinate axes. This is harder to interpret and should not be considered unless some particular meaning can be attached. Any rotation of the predictor space would reintroduce the interaction term and, as with the polynomials, we would not ordinarily want our model interpretation to depend on the particular basis for the predictors.

Thus hierarchically ordered sets of predictors imposes some restriction on the choice of predictors for a model.

10.2 Hypothesis Testing-Based Procedures

This is usually not a good idea but since these procedures are still prevalent, we describe them first before explaining the drawbacks.

Backward Elimination is the simplest of all variable selection procedures and can be easily implemented without special software. In situations where there is a complex hierarchy, backward elimination can be run manually while taking account of what variables are eligible for removal.

We start with all the predictors in the model and then remove the predictor with highest p-value greater than α_{crit}. Next refit the model and remove the remaining least significant predictor provided its p-value is greater than α_{crit}. Sooner or later, all "nonsignificant" predictors will be removed and the selection process will be complete.

The α_{crit} is sometimes called the "p-to-remove" and does not have to be 5%. If prediction performance is the goal, then a 15 to 20% cutoff may work best, although methods designed more directly for optimal prediction should be preferred.

Forward Selection just reverses the backward method. We start with no variables in the model and then for all predictors not in the model, we check their *p*-values if they are added to the model. We choose the one with lowest *p*-value less than α_{crit}. We continue until no new predictors can be added.

Stepwise Regression is a combination of backward elimination and forward selection. This addresses the situation where variables are added or removed early in the process and we want to change our mind about them later. At each stage a variable may be added or removed and there are several variations on exactly how this is done.

We illustrate backward elimination on some data on the 50 states from the 1970s. First we load the packages:

```
import pandas as pd
import numpy as np
import matplotlib.pyplot as plt
import scipy as sp
import statsmodels.api as sm
import statsmodels.formula.api as smf
import faraway.utils
```

The data were collected from U.S. Bureau of the Census. We will take life expectancy as the response and the remaining variables as predictors. We turn the State variable into the row index for convenience:

```
import faraway.datasets.statedata
statedata = faraway.datasets.statedata.load()
statedata.index = statedata['State']
statedata = statedata.drop('State',1)
statedata.head()
```

State	Population	Income	Illiteracy	LifeExp	Murder	HSGrad	Frost	Area
AL	3615	3624	2.1	69.05	15.1	41.3	20	50708
AK	365	6315	1.5	69.31	11.3	66.7	152	566432
AZ	2212	4530	1.8	70.55	7.8	58.1	15	113417
AR	2110	3378	1.9	70.66	10.1	39.9	65	51945
CA	21198	5114	1.1	71.71	10.3	62.6	20	156361

We fit the model:

```
lmod = smf.ols('LifeExp ~ Population + Income + Illiteracy + \
    Murder + HSGrad + Frost + Area', statedata).fit()
lmod.sumary()
```

	coefs	stderr	tvalues	pvalues
Intercept	70.943	1.748	40.59	0.0000
Population	0.000	0.000	1.77	0.0832
Income	-0.000	0.000	-0.09	0.9293
Illiteracy	0.034	0.366	0.09	0.9269
Murder	-0.301	0.047	-6.46	0.0000
HSGrad	0.049	0.023	2.10	0.0420
Frost	-0.006	0.003	-1.82	0.0752
Area	-0.000	0.000	-0.04	0.9649

n=50 p=8 Residual SD=0.745 R-squared=0.74

The signs of some of the coefficients match plausible expectations concerning how the predictors might affect the response. Higher murder rates decrease life expectancy as one might expect. Even so, some variables such as income, are not significant, contrary to what one might expect.

At each stage we remove the predictor with the largest *p*-value over 0.05.

```
lmod.pvalues.idxmax(), lmod.pvalues.max()
('Area', 0.9649)
```

Area is the first to go:

```
lmod = smf.ols('LifeExp ~ Population + Income + Illiteracy + \
    Murder + HSGrad + Frost', statedata).fit()
lmod.pvalues.idxmax(), lmod.pvalues.max()
('Illiteracy', 0.9340)
```

Now Illiteracy leaves the model:

```
lmod = smf.ols(
    'LifeExp ~ Population + Income + Murder + HSGrad + Frost',
    statedata).fit()
lmod.pvalues.idxmax(), lmod.pvalues.max()
('Income', 0.9153)
```

Followed by income.

```
lmod = smf.ols(
    'LifeExp ~ Population + Murder + HSGrad + Frost',
    statedata).fit()
lmod.pvalues.idxmax(), lmod.pvalues.max()
('Population', 0.0520)
```

Population might be the last variable to remove.

```
lmod = smf.ols(
    'LifeExp ~ Murder + HSGrad + Frost', statedata).fit()
lmod.sumary()
          coefs stderr tvalues pvalues
Intercept 71.036  0.983   72.25  0.0000
Murder    -0.283  0.037   -7.71  0.0000
HSGrad     0.050  0.015    3.29  0.0020
Frost     -0.007  0.002   -2.82  0.0070
```

n=50 p=4 Residual SD=0.743 R-squared=0.71

The final removal of the Population variable is a close call. We may want to consider including this variable if interpretation is made easier. Notice that the R^2 for the full model of 0.74 is reduced only slightly to 0.71 in the final model. Thus the removal of four predictors causes only a minor reduction in fit.

It is important to understand that the variables omitted from the model may still be related to the response. For example:

```
lmod = smf.ols(
    'LifeExp ~ Illiteracy + Murder + Frost', statedata).fit()
lmod.sumary()
           coefs stderr tvalues pvalues
Intercept  74.557  0.584  127.61  0.0000
Illiteracy -0.602  0.299   -2.01  0.0500
Murder     -0.280  0.043   -6.45  0.0000
Frost      -0.009  0.003   -2.94  0.0052
```

n=50 p=4 Residual SD=0.791 R-squared=0.67

We see that illiteracy does have some association with life expectancy. It is true that replacing illiteracy with high school graduation rate gives us a somewhat better fitting model, but it would be insufficient to conclude that illiteracy is not a variable of interest. Some might hope that this procedure could distinguish between important and unimportant predictors, but it does not do this in general.

Testing-based procedures are relatively cheap computationally and easy to understand, but they do have some drawbacks:

1. Because of the "one-at-a-time" nature of adding/dropping variables, it is possible to miss the "optimal" model.

2. The p-values used should not be treated too literally. There is so much multiple testing occurring that the validity is dubious. The removal of less significant predictors tends to increase the significance of the remaining predictors. This effect leads one to overstate the importance of the remaining predictors.

3. The procedures are not directly linked to final objectives of prediction or explanation and so may not really help solve the problem of interest. With any variable selection method, it is important to keep in mind that model selection cannot be divorced from the underlying purpose of the investigation. Variable selection tends to amplify the statistical significance of the variables that stay in the model. Variables that are dropped can still be correlated with the response. It would be wrong to say that these variables are unrelated to the response; it is just that they provide no additional explanatory effect beyond those variables already included in the model.

4. Stepwise variable selection tends to pick models that are smaller than desirable for prediction purposes. To give a simple example, consider the simple regression with just one predictor variable. Suppose that the slope for this predictor is not quite statistically significant. We might not have enough evidence to say that it is related to y, but it still might be better to use it for predictive purposes.

Except in simple cases where only a few models are compared or in highly structured heirarchical models, testing-based variable selection should not be used. We include it here because the method is still used, but should be discouraged. Hypothesis testing is best used for comparing just two models. One might stretch to a few more comparisons, but industrial scale usage destroys the justification for hypothesis testing without achieving a useful result.

The discussion above pertains mostly to prediction problems. For investigations where explanation is the goal, variable selection has a different role. Ideally one is focused on the potentially causal effect of a single predictor of interest. Other predictors are included in the model with the purpose of adjusting for their effect. We include these other predictors to ensure that the perceived effect of the predictor of interest cannot be explained by these other predictors. We might make some small gain in efficiency by removing unnecessary other predictors, but this would have little importance regarding the primary goal of determining the effect of the predictor of interest. Hence, in explanatory problems, the inclusion of other predictors is based on our assessment of their relationships with the other variables. We do not benefit from automated model selection in this situation.

10.3   Criterion-Based Procedures

If we have some idea about the purpose for which a model is intended, we might propose some measure of how well a given model meets that purpose. We would choose that model among those possible that optimizes that criterion.

We might have the idea to pick the model that is closest to the true relationship between the variables. With this in mind, we could pick a model g, parameterized by θ, that is close to the true model f. We could measure the distance between g and f by

$$I(f,g) = \int f(x) \log \left(\frac{f(x)}{g(x|\theta)} \right) dx$$

This is known as the Kullback-Leibler information (or distance). It will be positive except when $g = f$ when it will be zero. Unfortunately, it is impractical for direct implementation because we do not know f. If we knew f, we would simply use f. We can substitute in the MLE of θ and rearrange to obtain:

$$\hat{I}(f,g) = \int f(x) \log f(x)dx - \int f(x) \log g(x|\hat{\theta})dx$$

The first term is a constant that doesn't depend on the model we choose g. Akaike (1974) showed that $E\hat{I}(f,g)$ can be estimated by

$$- \log L(\hat{\theta}) + p + \texttt{constant}$$

where p is the number of parameters in the model and the constant depends on the unknown true model. For "historical" reasons, Akaike multiplied this by two to obtain "an information criterion" (AIC):

$$AIC = -2L(\hat{\theta}) + 2p$$

For linear regression models, the first term, $-2L(\hat{\theta})$ is equal to $n \log(RSS/n)$ plus another constant. Since the constants are the same for a given data set and assumed error distribution, they can be ignored for regression model comparisons on the same data. Additional care is necessary for other types of comparisons. See Burnham and Anderson (2002) for more detail.

We choose the model which minimizes the AIC. The first term in the AIC is based on RSS which is made smaller by improving the fit. Adding more predictors, provided they are not collinear, will achieve this but we are dissuaded from going too far with this by the second term, called the *penalty term*, $2p$. Hence, we see that AIC naturally provides a balance between fit and complexity in model selection.

An important idea in Statistics and Machine Learning is the *bias-variance trade-off*. In the context of model selection, a model which is too small will tend to have biased predictions because it is insufficiently flexible to represent the relationship between the predictors and the response. In contrast, a model which is too large will tend to overfit the response which will lead to greater variance in its prediction. The AIC represents a particular choice of this tradeoff.

Many other criteria have been proposed. One well-known alternative is the Bayes information criterion (BIC) which replaces the $2p$ term in the AIC with $p \log n$. BIC penalizes larger models more heavily and so will tend to prefer smaller models in comparison to AIC. AIC and BIC are often used as selection criteria for other types of models too. If we believe that there is a true model and we aim to choose it from a finite set of alternative models, BIC will choose the true model with probability approaching one as the sample size grows. On the other hand, suppose we do not believe in the true model but are willing to consider an increasing number of alternatives as our sample size grows. In this situation the AIC-selected model will tend to do as well as the best choice from those available. Unfortunately, these theories do not help much in deciding what to use in practice but there is some evidence to suggest that AIC is better for prediction problems.

Another straightforward alternative is the adjusted R^2 introduced in Section 2.9. The criterion favors a smaller RSS but without expending too many degrees of freedom. This follows the same idea as AIC or BIC but does not have the same amount of theoretical backing.

We demonstrate the use of AIC on the `state` dataset. We calculate the RSS for all possible subsets, returning the lowest value among all models with a fixed number of predictors:

```
import itertools
pcols = list(statedata.columns)
pcols.remove('LifeExp')
rss = np.empty(len(pcols) + 1)
rss[0] = np.sum(
    (statedata.LifeExp - np.mean(statedata.LifeExp))**2)
selvar = ['Null']
for k in range(1,len(pcols)+1):
    RSS = {}
    for variables in itertools.combinations(pcols, k):
        predictors = statedata.loc[:,list(variables)]
        predictors['Intercept'] = 1
        res = sm.OLS(statedata.LifeExp, predictors).fit()
        RSS[variables] = res.ssr
    rss[k] = min(RSS.values())
    selvar.append(min(RSS, key=RSS.get))
rss.round(3)
array([88.299, 34.461, 29.77 , 25.372, 23.308, 23.302, 23.298, 23.297])
```

These correspond to the models with the following predictors

```
['Null',
 ('Murder',),
 ('Murder', 'HSGrad'),
 ('Murder', 'HSGrad', 'Frost'),
 ('Population', 'Murder', 'HSGrad', 'Frost'),
 ('Population', 'Income', 'Murder', 'HSGrad', 'Frost'),
 ('Population', 'Income', 'Illiteracy', 'Murder', 'HSGrad',
  'Frost'),
 ('Population', 'Income', 'Illiteracy', 'Murder', 'HSGrad',
  'Frost', 'Area')]
```

Here we see that the best one predictor model uses Murder and so on. We compute and plot the AIC as seen in Figure 10.1:

```
aic = 50 * np.log(rss/50) + np.arange(1,9)*2
plt.plot(np.arange(1,8), aic[1:])
```

```
plt.xlabel('Number of Predictors')
plt.ylabel('AIC')
```

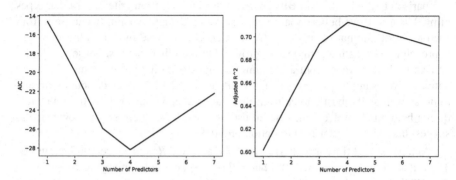

Figure 10.1   AIC and adjusted R^2 for models with varying numbers of predictors using the state data.

Alternatively, we can compute and plot the adjusted R^2 statistic:
```
adjr2 = 1 - (rss/(50 - np.arange(1,9)))/(rss[0]/49)
plt.plot(np.arange(1,8),adjr2[1:])
plt.xlabel('Number of Predictors')
plt.ylabel('Adjusted R^2')
```
The model chosen from these methods is the same — we pick the four predictor model using Population, Murder, HSGrad, Frost.

This method of computing the RSS for all possible subset models works fine when there are not too many predictors. The number of possible subsets is 2^q where q is the number of predictors. As q grows, this computation may become prohibitively time consuming. We can use more efficient programming techniques and algorithms but eventually we will need to give up on evaluating all possible subsets.

A less time-consuming method is to generate a sequence of models where there is one model for each number of predictors. We will not be able to exhaustively search all the possible subsets for a given number of predictors, but we might hope to generate a good choice less expensively. The scikit-learn package has a simple way to achieve this. The first step is to standardize the variables so they have a zero mean and unit standard deviation:
```
from sklearn.preprocessing import scale
scalstat = pd.DataFrame(scale(statedata), index=statedata.index,
    columns=statedata.columns)
```
The advantage of this rescaling is that the absolute size of the regression coefficient can be viewed as its relative importance within the model.
```
from sklearn import linear_model
reg = linear_model.LinearRegression(fit_intercept=False)
X = scalstat.drop('LifeExp',axis=1)
reg.fit(X, scalstat.LifeExp)
reg.coef_
array([ 0.17227607, -0.00998072,  0.0153566 , -0.82807917,  0.29440196,
       -0.22207364, -0.004693  ])
```
Note that the intercept is zero:

```
reg.intercept_
0.0
```

because we have standardized the data. We can see that the coefficient of Murder is the largest, while that of Area is the smallest indicating their relative importance in the model. The next step is to use *Recursive Feature Elimination* (RFE) which recursively eliminates the least important variable from the model, refits and repeats. The method is very similar to the backward elimination process presented earlier. The difference is that backward elimination is used to choose the model, while RFE merely gives a sequence of candidate models:

```
from sklearn.feature_selection import RFE
selector = RFE(reg, n_features_to_select=1)
selector = selector.fit(X, scalstat.LifeExp)
selector.ranking_
array([4, 6, 5, 1, 2, 3, 7])
```

These give the preference order for the predictors. We can see these with:

```
X.columns[np.argsort(selector.ranking_)].tolist()
['Murder', 'HSGrad', 'Frost', 'Population', 'Illiteracy', 'Income', 'Area']
```

As it happens, this generates the same sequence of candidate models as the exhaustive search. This is not guaranteed in general. We could now use AIC to select the model which, in this case, will make exactly the same choice. This example has such a small number of predictors that we would prefer the more exhaustive search described earlier, but in examples with more predictors, the method would be more useful.

Variable selection methods are sensitive to outliers and influential points. In this particular dataset, a check of the usual regression diagnostics reveals that Alaska has very high leverage. If we exclude this point and rerun the model selection procedures, you will find that a different set of predictors is selected. The selection methods are also sensitive to the transformation used on the variables. In this case, the Area and Population variables are quite skewed and we might apply a log transformation. If we do this and redo the model selection, yet another set of predictors is chosen. We recommend that analysts start by fitting a model with all the predictors and perform the regression diagnostics. Any unusual points or transformations should be dealt with before doing the model selection.

10.4 Sample Splitting

As we have seen, using the same data to fit and evaluate a model leads us to pick the most complex model from any set, as this will show the best apparent fit to the data. We can compensate for this by penalizing model complexity as seen in the AIC-based method. An alternative and more direct approach involves splitting the data into two parts. The first part, called the *training sample*, is used for fitting the model and the second, called the *testing sample*, is used for evaluating the model. For prediction problems, the true test of a model is in how well it predicts future observations. The test sample serves the role of this future sample. We use the predictor values in the test sample to predict the response and then compare this to the test sample responses.

We use the body fat density example from Section 4.2:

```
import faraway.datasets.fat
fat = faraway.datasets.fat.load()
```

We want to predict the fat density, given by the brozek variable, using up to 13 predictors which are age, height and weight along with 10 circumference measurements from the different parts of the body. There are 252 observations available.

The first step is to split the data into training and test samples:

```
n = len(fat)
np.random.seed(123)
ii = np.random.choice(n,n//3,replace=False)
testfat = fat.iloc[ii]
trainfat = fat.drop(ii)
```

We use a random split to ensure that there is no systematic difference between the two samples. We have set the random number seed for reproducibility. The outcome of the procedure does depend on the particular random split. One might be tempted to repeat the procedure with a different random split, but this would break the distinction between the training and test samples. We might consider putting the test sample in a virtual locked box to ensure that we do not use it for fitting models but only evaluating them.

There is no universal advice for the proportion of cases used for both samples. There is a tradeoff between estimating models well with a larger training sample and evaluating them well with a larger test sample. Experience suggests that the training sample should be larger. We have chosen to use two thirds of the data for this purpose.

The next step is to use the training data to generate a set of candidate models. In this example, we use the RFE method but any other reasonable method might be employed here. As in the previous example, we need to scale the data to use this method:

```
from sklearn.preprocessing import scale
scalfat = pd.DataFrame(scale(trainfat), columns=fat.columns)
```

Now we fit the full regression model with all 13 predictors:

```
from sklearn import linear_model
reg = linear_model.LinearRegression(fit_intercept=False)
prednames = ['age','weight','height','neck','chest','abdom','hip',
    'thigh','knee','ankle','biceps','forearm','wrist']
X = scalfat.loc[:,prednames]
reg.fit(X, scalfat.brozek)
```

We use the RFE method to generate a sequence of models, one of each size. The method returns a ranking of predictors:

```
from sklearn.feature_selection import RFE
selector = RFE(reg, n_features_to_select=1)
selector = selector.fit(X, scalfat.brozek)
selector.ranking_
array([10,  2, 11,  5,  8,  1,  6,  7, 12, 13,  9,  4,  3])
```

which corresponds to this ordering of the predictors:

```
np.array(prednames)[np.argsort(selector.ranking_)]
array(['abdom', 'weight', 'wrist', 'forearm', 'neck', 'hip', 'thigh',
    'chest', 'biceps', 'age', 'height', 'knee', 'ankle'], dtype='<U7')
```

We see that the single predictor model would use abdominal circumference, the two predictor model would add the weight and so on.

When evaluating models, we need a measure of performance. Our models will produce predicted values, ŷ, for the test sample which we will want to compare against the observed responses, y, in that sample. The root mean squared error (RMSE) is given by

$$RMSE = \sqrt{\frac{1}{n}\sum_{i=1}^{n}(y_i - \hat{y}_i)^2}$$

The RMSE is measured in the units of the response and might be considered the typical error in prediction. We define this with:

```
def rmse(x,y): return(np.sqrt(np.mean((x-y)**2)))
```

The first step is to create the design matrices for the training and test sets, along with vectors to store the RMSEs. In practice, we would only compute the RMSE for the test sample, but we perform the calculation for the training data also for comparative purposes. The null model with no predictors requires a separate calculation which we complete here.

```
Xtrain = trainfat.loc[:,prednames]
Xtest = testfat.loc[:,prednames]
pcols = Xtrain.shape[1]
prefs = np.argsort(selector.ranking_)
testpred = np.empty(pcols + 1)
testpred[0] = rmse(testfat.brozek, np.mean(trainfat.brozek))
trainpred = np.empty(pcols + 1)
trainpred[0] = rmse(trainfat.brozek, np.mean(trainfat.brozek))
```

Now we iterate over the set of candidate models. In each case, we fit the model using only the training data. For each model we fit, we make predictions using the test set and compare to the observed responses, using the RMSE to record the closeness. We do the same calculation for the training set just for comparison.

```
for k in range(1,pcols+1):
    reg = linear_model.LinearRegression(fit_intercept=True)
    reg.fit(Xtrain.iloc[:,prefs[0:k]], trainfat.brozek)
    ypred = reg.predict(Xtest.iloc[:,prefs[0:k]])
    testpred[k] = rmse(ypred,testfat.brozek)
    ypred = reg.predict(Xtrain.iloc[:,prefs[0:k]])
    trainpred[k] = rmse(ypred,trainfat.brozek)
```

We examine the RMSEs on the test sample:

```
testpred.round(3)
array([7.911, 4.986, 4.524, 4.44 , 4.451, 4.442, 4.436, 4.399, 4.421,
       4.408, 4.368, 4.409, 4.406, 4.41 ])
```

The smallest value occurs for the 12 predictor model. We could choose this model, but it is worthwhile to plot the RMSEs:

```
plt.plot(testpred,"k-",label="test")
plt.plot(trainpred,"k:",label="train")
plt.xlabel("No of predictors")
plt.ylabel("RMSE")
plt.legend()
```

In Figure 10.2, we see that the RMSEs for two or more predictors is very close. Although the performance might be marginally better for 12 predictors, we may be inclined to choose the two predictor model since this is much simpler and easier to use in practice. Also bear in mind the random nature of the train-test split which might produce a somewhat different outcome if repeated with a new split. This also inclines us towards the simplest model that we can justify.

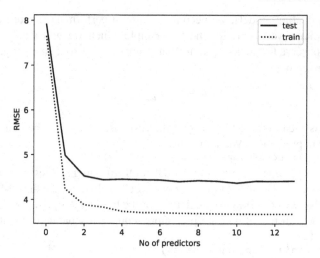

Figure 10.2   Model performance on the testing and training samples for the fat density data.

The RMSE values for the training sample are seen to be consistently substantially lower than those seen in the test sample. This illustrates the overconfidence that would result from using the same data for both fitting and evaluation.

The usage of sample splitting varies in practice according to the needs and motivations of the individuals involved. In a low-trust environment, unscrupulous analysts may consider a very large number of potential models and use the same data to both fit and choose the model. They conceal their search process and report only the final model and its purported performance. Unfortunately, it is not unheard of for researchers to gain acceptance for academic papers or consultants to sell models to industry clients on this basis. The miscreants receive their reward well before the truth behind their over-hyped models is revealed by fresh data. It is also easy enough to engage in much the same behaviour through ignorance rather than malpractice.

To protect against this malpractice, the data is split by some trusted guardian and only the training sample is given to the analysts. Analysts are invited to develop and submit a model based only on the training sample. The guardian then makes the predictions on the test sample and reports the result. The RMSE (or other measure of performance) on the test sample represents an unbiased estimate of its performance on future data. This is the mechanism used in prediction competitions like M4 Makridakis, Spiliotis, and Assimakopoulos (2018).

Viewed from this perspective, we have already cheated a little in the example above since we have submitted not one model but fourteen for evaluation. This will be fine for the purposes of choosing the best model, but the observed RMSE will now be biased downwards since we have chosen the smallest from those available. Since our set of candidate models is small, the bias will not be large but if we were to substantially expand the set of possibilities, the problem would become worse.

Sample splitting is clearly worthwhile strictly for the purposes of model evaluation. We pay for the honest evaluation with what may be a poorer choice of model with more variable parameter estimates. Now suppose we will use the model for own purposes. If we have skin in the game, we will only be cheating ourselves if we overfit the model. In these circumstances (and for the unimpeachably honest), sample splitting would make less sense. We might prefer to avoid splitting in this situation.

Returning to our fat density example, we will want to propose a model for predicting future responses. We have chosen our model and now we need to estimate the parameters of that model. We have choice whether to use just the training sample or the full data. If we need to convince others regarding the potential performance of our full model, we should restrict ourselves to the training data. On the other hand, if we want to obtain the best predictions including accurate estimates of uncertainty, it is usually best to use the full data. A full explanation of this can be found in Faraway (2016).

Sample splitters should understand the reason they are using the method. They also need to consider the degree of trust they expect from others in evaluating the outcome. They also need to remember that sample splitting protects them only from overconfidence and not the other pitfalls of prediction described in Section 4.4.

10.5 Crossvalidation

Sample splitting employs a single random split. For model selection purposes, we are not using the data efficiently when we use only part for estimation and part for validation. Crossvalidation is motivated by a desire to make the process more efficient with the data and reduce the variation caused by the random splitting.

To crossvalidate, we divide the data randomly into k parts as equally as possible. We use $k - 1$ parts to fit the model and the remaining part to evaluate it. We repeat the process k times, in each instance leaving out a different part of the data. We average the performance measure across these splits to obtain an overall measure of performance of the model.

Let's see how this works with the fat density example. First, we scale the full sample and extract the design matrix. We fit a regression model with all the predictors to initialize the process.

```
scalfat = pd.DataFrame(scale(fat), columns=fat.columns)
X = scalfat.loc[:,prednames]
reg = linear_model.LinearRegression(fit_intercept=False)
reg.fit(X, scalfat.brozek)
```

Now we load the crossvalidated version of the RFE method as the RFECV function from scikit-learn. We choose $k = 10$. We also specify the measure of performance using a scoring function. We choose the negative mean squared error. RFECV wants a criterion where more positive values are better which explains the negative in the specification. The MSE and RMSE will make the same choices so we can make with what RFECV provides here. By default, RFECV will work through the same sequence of candidate models as the RFE method. In this case, the predictor with the smallest parameter estimate is eliminated at each stage (which is why we needed to scale the data).

```
from sklearn.feature_selection import RFECV
selector = RFECV(reg, step=1, cv=10,
    scoring='neg_mean_squared_error')
selector = selector.fit(X, scalfat.brozek)
```

We can plot the scores for each model as seen in Figure 10.3.

```
plt.plot(np.arange(1,14),-selector.grid_scores_)
plt.xlabel("No. of Predictors")
plt.ylabel("MSE")
```

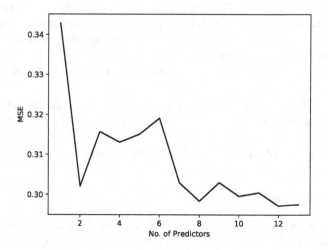

Figure 10.3   Crossvalidated model performance for the fat density data.

We see that the model with 12 predictors produces the smallest MSE. Even so, we might prefer the two predictor choice on the grounds that it has almost as good a performance while being substantially simpler. The chosen model is given by:

```
selector.ranking_
```
array([1, 1, 1, 1, 1, 1, 1, 1, 2, 1, 1, 1, 1])

All the chosen predictors are denoted by one with any remaining predictors ranked in order of preference. In this case, only the knee circumference is discarded.

The lack of convexity in the scores as a function of the predictors is dissappointing but the crossvalidated split is random and there is no guarantee of actual convexity anyway.

We chose $k = 10$ which is a good compromise between the computational expense of larger values and the variation seen in smaller choices. At one extreme, there is the possibility of *leave-out-one* crossvalidation where $k = n$. This has the advantage of avoiding the randomness induced by smaller values of k. On the other hand, it is very computational expensive in general since the model needs to be fit n times (although there are sometimes shortcuts available). Furthermore, it is not always effective as a means of model selection in all scenarios. Empirical evidence suggests choices like $k = 10$ are better.

Harrell (2015) has extensive advice on model validation using sample splitting and crossvalidation. Bootstrap methods are also strongly recommended. Steyerberg (2009) also provides advice on model selection and validation in clinical applications.

10.6 Summary

Model selection is a means to an end and not an end itself. When the aim is to construct a model that explains the relationships in the data, automatic variable selection is not helpful. Even when prediction is the goal, some human intelligence is sometimes necessary in supervising the process.

Hypothesis testing-based methods can be useful when choosing within a small set of possible models or when models are hierarchically structured. When considering a larger set of models, testing-based methods use a restricted search through the model space and use a dubious method for choosing between models when tests are repeated many times. Criterion-based methods typically involve a wider search and compare models in a preferable manner. Sample splitting methods provide a flexible alternative and are helpful particularly in low-trust situations. They are also inefficient methods of model selection particularly when the sample size is not large. Cross-validation addresses some of the inefficiency and instability of sample splitting methods. For data which are not extremely large in size and where the model space is not especially large, our preference is for criterion-based methods such as AIC.

Accept the possibility that several models may be suggested which fit about as well as each other. If this happens, consider:

1. Do the models have similar qualitative consequences?

2. Do they make similar predictions?

3. What is the cost of measuring the predictors?

4. Which has the best diagnostics?

If you find models that seem roughly comparable, but lead to quite different conclusions, then it is clear that the data cannot answer the question of interest unambiguously. Be alert to the possibility that a model contradictory to the tentative conclusions might be out there.

Exercises

1. Use the `prostate` data with `lpsa` as the response and the other variables as predictors.

   (a) Fit the full model and determine which predictors have p-values less than 5%. Now use the backward elimination method to choose a model based on a 5% cutoff for the p-value. Do the two procedures pick the same set of predictors? Is this guaranteed to happen for all datasets?

   (b) Use Recursive Feature Elimination to generate a sequence of models of decreasing size. Does this generate the same sequence of models as backward elimination? Explain why these might be different in general.

(c) Consider all possible subsets of the full model and find the one with the best value of AIC. Is the set of models of each size the same as that chosen by RFE? Would the best model from the RFE generated sequence produce the same choice as from the exhaustive search?

(d) What model is chosen when adjusted R^2 is the criterion?

(e) Explain why sample splitting might be problematic for this dataset.

2. Using the `teengamb` dataset with `gamble` as the response and the other variables as predictors. Focus on the effect of sex on the response and include this predictor in all models. There are eight possible models that include all, some, or none of the other three predictors. Fit all these models and report on the coefficient and significance of sex in each case as well as the adjusted R^2 and AIC values.

(a) Does the significance of sex vary according to the model chosen?

(b) Which model is chosen by the Adjusted R^2 criterion? By AIC?

(c) Four of the models have distinctly better fits from the other four. What distinguishes these models?

(d) Make a plot of income against amount gambled which distinguishes male from female subjects. Discuss how the amount gambled by males and females varies according to income.

3. Using the `divusa` dataset, fit a model with `divorce` as the response, and the other variables as predictors with the exception of `year`.

(a) Which predictors are not significant at the 5% level?

(b) Remove the insignificant predictors and perform an F-test to compare the reduced model to the original model.

(c) Consider all models with 4 predictors from the original set of 5. For each model (including the full model), compute the adjusted R^2, AIC, the predicted value of the response at the maximum observed values of the predictors along with standard error for this prediction. Comment on the results.

4. Using the `trees` data, fit a model with `log(Volume)` as the response and a second-order polynomial (including the interaction term) in `log(Girth)` and `log(Height)`.

(a) What terms are significant in the quadratic model? Does the model fit badly?

(b) Consider smaller models for the data.

5. Using the `stackloss` data, fit a model with `stack.loss` as the response and the other three variables as predictors. You can get the stackloss data using `{stackloss = sm.datasets.stackloss.load_pandas().data}`

(a) Fit the model using least squares. Use hypothesis testing to select a model.

(b) Now fit the model using least absolute deviations. Which model would you select if this estimation method is used?

(c) Would it be possible to use an estimation method other than LS with the AIC-criterion based choice of model? Could we use the sample splitting or CV approach?

6. Use the `seatpos` data with `hipcenter` as the response.

   (a) Use the AIC criterion to choose the best model.

   (b) Now add an alternative sequence of -1cm and +1cm to the response using `np.tile([-1,1],19)`. Repeat the model search process. Is the generated sequence of best models of each size the same as previously?

7. In this question, we select the amount of B-spline smoothing to fit a function.

   (a) Generate and plot some simulated data as described at the beginning of Section 9.5.

   (b) For each number of knots up to a maximum of 25, fit the B-spline model with that number of equally-spaced knots. For each model, calculate and save the adjusted R^2 and AIC.

   (c) What number of knots maximizes the adjusted R^2?

   (d) What number of knots minimizes the AIC?

8. Use the `newhamp` data as described in Section 5.4.

   (a) Discard the first four variables: `votesys`, `Obama`, `Clinton`, `dem`.

   (b) Split the data randomly into 2/3 training and 1/3 testing.

   (c) Consider pObama as the response and the remaining variables as predictors. Use the method described in Section 10.4 to select a model.

   (d) Now use the crossvalidation method as demonstrated in the chapter to select a model.

Chapter 11

Shrinkage Methods

Sometimes it can seem like you have too many predictors. If you use them all in your regression model, problems can arise. Explanation can be difficult due to collinearity as we saw in Section 7.3. Prediction performance can also be degraded by using too many predictors. But more predictors should mean more information. That should be helpful. In this chapter, we will look at four methods that allow us to *shrink* this additional information into a more useful form. You will see later why we have used the word "shrink".

11.1 Principal Components

Principal components analysis (PCA) is a popular method for finding low-dimensional linear structure in higher dimensional data. It has a range of purposes, but let's see how it can be helpful in regression problems.

Designed experiments typically have predictors which are mutually orthogonal. This makes the fitting and interpretation of these models simpler. For observational data, predictors are often substantially correlated. It would be nice if we could transform these predictors to orthogonality, as it could make interpretation easier.

For example, consider the dimensions of the human body as measured in a study on 252 men as described in Johnson (1996) as seen partially in Figure 11.1. We analyzed this data in Section 4.2 where more details about motivation behind the data collection can be found. First we load the packages:

```
import pandas as pd
import numpy as np
import scipy as sp
import statsmodels.api as sm
import statsmodels.formula.api as smf
import faraway.utils
import matplotlib.pyplot as plt
```

and now the data:

```
import faraway.datasets.fat
fat = faraway.datasets.fat.load()
```

Some plots of the data are helpful, for example:

```
plt.scatter(fat.knee, fat.neck)
plt.xlabel("Knee")
plt.ylabel("Neck")
plt.scatter(fat.thigh, fat.chest)
plt.xlabel("Thigh")
plt.ylabel("Chest")
plt.scatter(fat.wrist, fat.hip)
plt.xlabel("Wrist")
```

```
plt.ylabel("Hip")
```

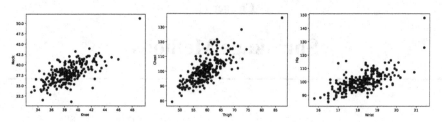

Figure 11.1   Circumferences in centimeters for body dimensions of 252 men.

We see that the body circumferences are strongly correlated. Although we may have many predictors, there may be less information than the number of predictors might suggest. PCA aims to discover this lower dimension of variability in higher dimensional data.

PCA can be understood as a particular rotation of the data. We want to rotate the data around its mean value. To this end, we center the matrix of predictors X by subtracting the mean for each variable so that the columns of X have mean zero. We use an X that does not include a column of ones for an intercept term.

The PC decomposition can be computed intuitively as follows:

1. Find the direction of greatest variation in the data. We can do this by finding a vector u_1 such that var (Xu_1) is maximized subject to $u_1^T u_1 = 1$. The u_1 describes this axis of greatest variation. We call $z_1 = Xu_1$ the first principal component (PC).

2. The data has now been rotated so that the direction of greatest variation coincides with the axis of the first principal component. With this axis fixed, we find a direction orthogonal to this which has the greatest remaining variation. In other words, we find u_2 such that var (Xu_2) is maximized subject to $u_1^T u_2 = 0$ and $u_2^T u_2 = 1$. We call $z_2 = Xu_2$ the second PC.

3. We keep finding directions of greatest variation orthogonal to those directions we have already found. For low dimensions, the process can be iterated until all dimensions have been exhausted. For high dimensional problems, we can stop when the remaining variation is negligible.

We write $z_i = Xu_i$ and the z_i are the PCs. We can gather the terms in the matrix form $Z = XU$ where Z and U have columns z_i and u_i, respectively. U is called the rotation matrix. We can think of Z as a version of the data rotated to orthogonality.

In practice, it is more efficient to use the singular value decomposition (SVD) to compute the PCA. The SVD has also been used to compute least squares estimates as discussed in Section 2.7.

We consider only the circumference measurements in the fat dataset and compute the principal components decomposition:

```
from sklearn.decomposition import PCA
pca = PCA()
cfat = fat.iloc[:,8:]
pca.fit(cfat)
```

We start by looking at the standard deviations of the principal components, $SD(z_i)$:

```
np.sqrt(pca.explained_variance_).round(2)
array([15.99,  4.07,  2.97,  2.  ,  1.69,  1.5 ,  1.3 ,  1.25,  1.11, 0.53])
```

Sometimes, it can also be helpful to consider what proportion each principal component contributes to the total variation. We can check this with:

```
pca.explained_variance_ratio_.round(3)
array([0.867, 0.056, 0.03 , 0.014, 0.01 , 0.008, 0.006, 0.005, 0.004, 0.001])
```

We can see that the first PC explains a large fraction of the total variation. The second PC explains much less and later PCs hardly anything. Given that these measurements describe the shape of a human, we are not surprised. We may vary in size but we are all similar in shape. If we were to add some animals or aliens into our data, we might expect more variation in the other PCs.

Instead of ten variables, we could use just a single variable, formed by a linear combination described by the first PC, which would represent the ten-dimensional data quite well.

The first column of the rotation matrix, u_1, is a linear combination describing the first principal component given by rot[0,:] which we print neatly on one row with:

```
rot = pca.components_
pd.DataFrame(rot[0,:],index=cfat.columns).round(3).T
```

neck	chest	abdom	hip	thigh	knee	ankle	biceps	forearm	wrist
0.122	0.502	0.658	0.42	0.28	0.121	0.056	0.145	0.074	0.039

We see that the chest, abdomen, hip and thigh measures dominate the first principal component. However, the reason for this may simply be that these measures are larger and so more variable than the wrist or ankle circumferences. We might prefer to scale the variables for size by converting to standard units, that is, subtracting the mean and dividing by the standard deviation. We can achieve this as follows:

```
from sklearn.preprocessing import scale
scalfat = pd.DataFrame(scale(cfat))
pcac = PCA()
pcac.fit(scalfat)
pcac.explained_variance_ratio_.round(2)
array([0.702, 0.073, 0.067, 0.049, 0.03 , 0.028, 0.02 , 0.016, 0.008, 0.006])
```

We can see that, after scaling, the proportion of variablity explained by the first component drops to 70.2%. The remaining variation is more evenly spread over the other components. Here is the first principal component:

```
rot = pcac.components_
pd.DataFrame(rot[0,:],index=cfat.columns).round(3).T
```

neck	chest	abdom	hip	thigh	knee	ankle	biceps	forearm	wrist
0.327	0.339	0.334	0.348	0.333	0.329	0.247	0.322	0.27	0.299

It has very similar coefficients for all the variables. It is useful to interpret this as "overall size" since it is roughly proportional to a mean across all these (standardized) variables. One interpretation of this dominant first principal component is that body shapes in men are mostly proportional. Bigger men tend to be just larger all-round versions of smaller men.

The other principal components describe how the data vary in ways orthogonal to this first PC. For example, we might look at the second principal component:

```
pd.DataFrame(rot[1,:],index=cfat.columns).round(3).T
```

neck	chest	abdom	hip	thigh	knee	ankle	biceps	forearm	wrist
-0.024	0.384	0.384	-0.509	-0.6	-0.175	-0.115	-0.183	-0.088	-0.014

which is roughly a contrast between the body center measures of chest, abdomen, hip and thigh circumferences against the extremities of forearm, wrist and ankle measures. This could be viewed as a relative measure of where the body is carrying its weight. This represents only 7.3% of the total variation so it is not substantial but is the largest component of variation after the first PC.

So far we have not tried to link the predictors to the response in a regression model. We have considered only the predictors so far. Indeed, PCA is widely used in other applications — see Joliffe (2002) for a book length treatment. From what we have seen so far in the fat density example, the predictors are highly correlated and most of their variation can be viewed as a measure of overall body size. We would expect some problems with collinearity and we might wonder whether we need all these predictors, but we need to move on to the regression before taking any decisions.

Robust PCA

Like variances, principal components analysis can be very sensitive to outliers so it is essential to check for these. In addition to the usual graphical checks of the data, it is worth checking for outliers in higher dimensions. Such outliers can be hard to find. For example, consider someone with a weight of 50 kg and a height of 2 m. Neither value is individually exceptional, but the combination would be very unlikely and possibly an error.

The minimum covariance determinant method of Rousseeuw and Driessen (1999) is a highly robust estimator of multivariate location and scatter. We specify some number h which is a majority of the n observations that we believe are reliable. We then find h observations whose covariance matrix has the lowest determinant (or roughly speaking, overall variance). The EllipticEnvelope from scikit-learn implements this procedure. By default, 10% of the observations are assumed to be unreliable.

```
from sklearn.covariance import EllipticEnvelope
ee = EllipticEnvelope()
ee.fit(cfat)
```

We could simply take the covariance of this central subset of the data as the input for the PCA, but we may wish to identify (and perhaps discard) the outliers. We might look for points which are far from the mean but this needs to take account of how the data varies. Mahalanobis distance is a measure of the distance of a point from the mean that adjusts for the correlation in the data. It is defined as

$$d_i^2 = (x - \mu)^T \Sigma^{-1} (x - \mu)$$

where μ is a measure of center and Σ is a measure of covariance. We compute these and make a half-normal plot of the d_i as seen in seen in Figure 11.2.

```
md = np.sqrt(ee.mahalanobis(cfat))
n=len(md)
ix = np.arange(1,n+1)
halfq = sp.stats.norm.ppf((n+ix)/(2*n+1)),
plt.scatter(halfq, np.sort(md))
plt.xlabel(r'$\chi^2$ quantiles')
```

```
plt.ylabel('Mahalanobis distances')
```

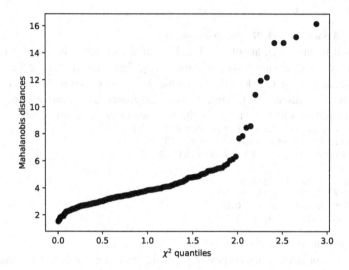

Figure 11.2   Half-normal plot of the Mahalanobis distances for the body circumference data.

We see that there are some outliers and that we can investigate the sensitivity of the PCA to these values by re-analyzing the data after removing these points. In this example, the second PC shows more of a change.

PCA for regression

Now let's see how we can use PCA for regression. We might have a model y ~ X. We replace this by y ~ Z where typically we use only the first few columns of Z. This is known as *principal components regression* or PCR. The technique is used in two distinct ways for explanation and prediction purposes.

When the goal of the regression is to find simple, well-fitting and understandable models for the response, PCR *may* help. The PCs are linear combinations of the predictors. For example, the purpose in collecting the body fat data was to model the percentage of body fat described by the response, brozek. We analyzed this data before in Section 4.2, but we are using fewer predictors here. Here is the model output where we use all ten circumference predictors:

```
xmat = sm.add_constant(cfat)
lmod = sm.OLS(fat.brozek, xmat).fit()
lmod.sumary()
       coefs stderr tvalues pvalues
const  7.229  6.214    1.16  0.2459
neck  -0.582  0.209   -2.79  0.0057
chest -0.091  0.085   -1.06  0.2887
abdom  0.960  0.072   13.41  0.0000
hip   -0.391  0.113   -3.47  0.0006
thigh  0.134  0.125    1.07  0.2855
knee  -0.094  0.212   -0.44  0.6583
ankle  0.004  0.203    0.02  0.9834
```

```
biceps    0.111  0.159    0.70  0.4853
forearm   0.345  0.186    1.86  0.0645
wrist    -1.353  0.471   -2.87  0.0045
```

```
n=252 p=11 Residual SD=4.071 R-squared=0.74
```

It is difficult to say much about which factors might influence body fat percentage because there are clear indications of collinearity. The signs of the coefficients and their significance vary in a less than credible way. Why would abdomen circumference have a positive effect, while hip circumference has negative effect? Now consider the output where we use only the first two principal components:

```
pcscores = pca.fit_transform(scale(cfat))
xmat = sm.add_constant(pcscores[:,:2])
lmod = sm.OLS(fat.brozek, xmat).fit()
lmod.sumary()
        coefs stderr tvalues pvalues
const  18.938  0.329   57.54  0.0000
x1      1.838  0.124   14.80  0.0000
x2     -3.543  0.386   -9.18  0.0000
```

```
n=252 p=3 Residual SD=5.225 R-squared=0.55
```

We have lost some explanatory power in going from ten down to two predictors. But these two predictors are now orthogonal, meaning we can now interpret these without collinearity worries. As previously discussed, the first PC can be viewed as a measure of overall size. We can see this is associated with higher body fat. The second PC shows a negative association, meaning that men who carry relatively more of their substance in their extremities tend to be leaner. These would tend to be men who are more muscular so this result accords with what one might expect. So the PCR here has allowed us a meaningful explanation, whereas the full predictor regression was opaque.

One objection to the previous analysis is that the two PCs still use all ten predictors, so there has been no saving in the number of variables needed to model the response. Furthermore, we must rely on our subjective interpretation of the meaning of the PCs. To answer this, one idea is to take a few representative predictors based on the largest coefficients seen in the PCs. For example, we might pick out the abdomen circumference to represent the first PC and the difference between abdomen and ankle circumference for the second PC. The latter pair show the largest coefficients in absolute value (-0.40 and 0.62) for the second PC. Abdomen does not quite have the largest coefficient in the first PC (hip does), but it's a close choice and means we need only two predictors in the following model. We need to scale the predictors so that ankle and abdomen are on the standard units scale.

```
xmat = pd.concat([scalfat.iloc[:,2],
    scalfat.iloc[:,6] - scalfat.iloc[:,2]], axis=1)
xmat.columns = ['overall','muscle']
xmat = sm.add_constant(xmat)
lmod = sm.OLS(fat.brozek, xmat).fit()
lmod.sumary()
          coef stderr tvalues pvalues
const   18.938  0.279   67.79  0.0000
overall  5.751  0.328   17.55  0.0000
muscle  -0.993  0.313   -3.17  0.0017
```

```
n=252 p=3 Residual SD=4.435 R-squared=0.68
```

We have a simple model that fits almost as well as the ten-predictor model. We have given names to the two predictors. We can interpret it similarly to the previous model but it is easier to explain to others. Future studies might be done more cheaply because we might only need these two measures.

However, for this explanatory use of PCR to work, we typically need the predictors to measure quantities for which linear combinations are interpretable — usually the predictors would need to have the same units. So if we had used the age and weight variables found in the `fat` data example, it would have been far more difficult to interpret the linear combinations. Even in the homogeneous predictor case, we need some luck and imagination to get interpretable PCs. These requirements restrict the utility of PCR for explanatory purposes. It is worth trying but you might not get anything useful from it.

We can sometimes make better predictions with a small number of principal components in Z than a much larger number of variables in X. Success requires that we make a good choice of the number of components. To illustrate the use of PCR and the other shrinkage methods in this chapter, we will use a set of data where the emphasis is on prediction but the explanatory aspects of the methods can be useful in gaining intuition about the structure of the data. A Tecator Infratec Food and Feed Analyzer working in the wavelength range of 850 to 1050 nm by the near-infrared transmission (NIT) principle was used to collect data on samples of finely chopped pure meat and 215 samples were measured. For each sample, the fat content was measured along with a 100-channel spectrum of absorbances. Since determining the fat content via analytical chemistry is time consuming, we would like to build a model to predict the fat content of new samples using the 100 absorbances which can be measured more easily. See Thodberg (1993) for more details.

The true performance of any model is hard to determine based on just the fit to the available data. We need to see how well the model does on new data not used in the construction of the model. For this reason, we will partition the data into two parts — a *training sample* consisting of the first 172 observations that we will use to build and estimate the models and a *testing sample* of the remaining 43 observations. We will use the test sample only to evaluate the performance of our models and not to select them.

We load in the data and make the split into training and testing samples: Let's start with the least squares fit:

```
import faraway.datasets.meatspec
meatspec = faraway.datasets.meatspec.load()
trainmeat = meatspec.iloc[:172,]
testmeat = meatspec.iloc[173:,]
```

A linear model with all the predictors is a good place to start. We use the `LinearRegression` function from `linear_model` module in `scikit-learn`. The function has far fewer features than those in `statsmodels`, but it works consistently with other functions in `scikit-learn` and so it is more convenient to use it here. In this function, the `score` is the R^2.

```
from sklearn import linear_model
```

```
fullreg = linear_model.LinearRegression(fit_intercept=True)
Xtrain = trainmeat.drop('fat',axis=1)
fullreg.fit(Xtrain, trainmeat.fat)
fullreg.score(Xtrain, trainmeat.fat)
0.99702
```

We see that the fit of this model is already very good in terms of R^2. How well does this model do in predicting the observations in the test sample? We need a measure of performance — we use root mean square error (RMSE):

$$\sqrt{(\sum_{i=1}^{n} (\hat{y}_i - y_i)^2 / n)}$$

The MSE can be found in the metrics module.
```
from sklearn.metrics import mean_squared_error
```
We find the RMSE for the training sample:
```
np.sqrt(mean_squared_error(fullreg.predict(Xtrain),trainmeat.fat))
```
We prefer RMSE to MSE because RMSE is measured in the units of the response which makes it easier to interpret.
```
0.69032
```
For the test sample:
```
Xtest = testmeat.drop('fat',axis=1)
np.sqrt(mean_squared_error(fullreg.predict(Xtest),testmeat.fat))
3.85908
```
We see that the performance is much worse for the test sample. This is not unusual, as the fit to the data we have almost always gives an over-optimistic sense of how well the model will do with future data. In this case, the actual error is about five times greater than the model itself suggests.

Now, it is quite likely that not all 100 predictors are necessary to make a good prediction. Let's take a look at the predictors as seen in the left panel of Figure 11.3:
```
frequency = np.arange(0,100)
plt.plot(frequency,Xtrain.T,alpha=0.15)
plt.xlabel("Frequency")
```
We can see that there is a high degree of correlation between adjacent frequencies. This suggests that we might do without many of them. We use the recursive feature elimination method with crossvalidation to select the number of predictors:
```
from sklearn.feature_selection import RFECV
redreg = linear_model.LinearRegression(fit_intercept=True)
redreg.fit(Xtrain, trainmeat.fat)
selector = RFECV(redreg, step=1, cv=10)
selector = selector.fit(Xtrain, trainmeat.fat)
Xred = Xtrain.iloc[:,selector.support_]
Xred.shape
(172, 19)
```
Only 19 predictors are used. Now compute the RMSE on the training set:
```
redreg.fit(Xred,trainmeat.fat)
np.sqrt(mean_squared_error(redreg.predict(Xred),trainmeat.fat))
2.74545
```
and now on the test set:
```
Xredtest = Xtest.iloc[:,selector.support_]
np.sqrt(mean_squared_error(redreg.predict(Xredtest),testmeat.fat))
2.51918
```

The model selection step removed 81 variables. Of course, the nominal fit got a little worse as it always will when predictors are removed, but the actual performance improved from 3.86 to 2.52.

Now let's compute the PCA on the training sample predictors. We examine the standard deviations of the ten largest principal components:

```
from sklearn.decomposition import PCA
pca = PCA()
pca.fit(Xtrain)
np.sqrt(pca.explained_variance_).round(2)[:10]
array([5.06, 0.51, 0.28, 0.17, 0.04, 0.02, 0.01, 0.01, 0.  , 0.  ])
```

We see that the first PC accounts for about ten times more variation than the second. The contribution drops off sharply. This suggests that most of the variation in the predictors can be explained with just a few dimensions.

The linear combinations u_i (or *loadings*) can be found in the rows of the components part of the PCA fitted object. We plot these vectors against the frequency in Figure 11.3 using:

```
plt.plot(frequency, pca.components_[0,:],'k-',label="PC1")
plt.plot(frequency, pca.components_[1,:],'k:',label="PC2")
plt.plot(frequency, pca.components_[2,:],'k--',label="PC3")
plt.legend()
plt.xlabel("Frequency")
plt.ylabel("Coefficient")
```

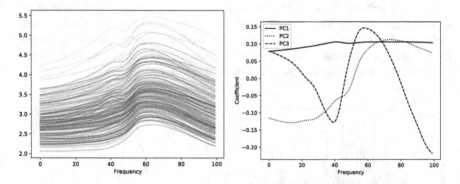

Figure 11.3   The spectra for training sample are shown on the left. On the right, we see eigenvectors (loadings) for the PCA of the meat spectrometer data.

These vectors represent the linear combinations of the predictors that generate the PCs. We see that the first PC comes from an almost constant combination of the frequencies. It measures whether the predictors are generally large or small. The second PC represents a contrast between the higher and lower frequencies. The third is more difficult to interpret. It is sometimes possible, as in this example, to give some meaning to the PCs. This is typically a matter of intuitive interpretation. Sometimes, no interpretation can be found and we must be satisfied with possibility of improved prediction.

Let's use the first four PCs to predict the response:

```
pcscores = pca.fit_transform(Xtrain)
```

```
pc4reg = linear_model.LinearRegression(fit_intercept=True)
pc4 = pcscores[:,:4]
pc4reg.fit(pc4,trainmeat.fat)
np.sqrt(mean_squared_error(pc4reg.predict(pc4),trainmeat.fat))
4.06474
```

We do not expect as good a fit using only four variables instead of the 100. Even so, considering that, the fit is not much worse than the results from much bigger models.

PCR is an example of *shrinkage* estimation. Let's see where the name comes from. We plot the 100 slope coefficients for the full least squares fit:

```
plt.plot(frequency, fullreg.coef_)
plt.xlabel("Frequency")
plt.ylabel("Coefficient")
```

which is shown in the left panel of Figure 11.4. We see that the coefficients' range is in the thousands and that the adjacent coefficients can be very different. This is perhaps surprising because one might expect that adjacent frequencies might have a very similar effect on the response. We plot the coefficients from the four-component fit in the panel on the right of Figure 11.4.

```
pceff = np.dot(pca.components_[:4,].T, pc4reg.coef_)
plt.plot(frequency, pceff)
plt.xlabel("Frequency")
plt.ylabel("Coefficient")
```

Here we see that the range of these coefficients is much smaller than the thousands seen for the ordinary least squares fit. Instead of wildly varying coefficients in the least squares case, we have a more stable result. This is why the effect is known as shrinkage. Furthermore, there is smoothness between adjacent frequencies as knowledge of the science behind this data might suggest. We also see that the prediction is constructed mostly using the lower half of the frequencies.

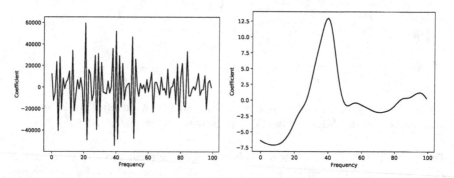

Figure 11.4   Coefficients for the least squares fit on the left and for the PCR with four components on the right.

Why use four PCs here and not some other number? The standard advice for choosing the number of PCs to represent the variation in X is to choose the number beyond which all the PC standard deviations are relatively small. We could make a case for using only the first PC, but look at the sequence of SDs of PCs:

$5.06, 0.51, 0.28, 0.17, 0.04, \ldots$ There is another identifiable "elbow" at five indicating the choice of four PCs.

Now let's see how well the test sample is predicted. We need to center the test predictors by the mean from the training set and then use the $Z = XU$ formula to compute the scores for the prediction.

```
rotX = (Xtest - pca.mean_) @ pca.components_[:4,].T
np.sqrt(mean_squared_error(pc4reg.predict(rotX),testmeat.fat))
4.52586
```

which is not at all impressive. It turns out that we can do better by using more PCs — we figure out how many would give the best result on the test sample:

```
maxcomp = 50
ncomp = np.arange(1,maxcomp+1)
rmsep = np.empty(maxcomp)
pcrmod = linear_model.LinearRegression(fit_intercept=True)
for icomp in ncomp:
    rotX = (Xtest - pca.mean_) @ pca.components_[:icomp,].T
    pcsi = pcscores[:,:icomp]
    pcrmod.fit(pcsi,trainmeat.fat)
    rmsep[icomp-1] = np.sqrt(mean_squared_error(
        pcrmod.predict(rotX),testmeat.fat))
plt.plot(ncomp, rmsep)
plt.ylabel("RMSE")
plt.xlabel("Number of components")
```

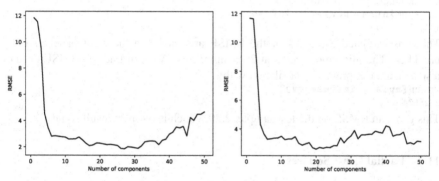

Figure 11.5   RMS for the test sample on the left and RMS estimated using CV for varying numbers of PCs on the right.

The plot of the RMSE is seen in Figure 11.5. Here is the best choice:

```
np.argmin(rmsep)+1, min(rmsep)
(27, 1.8687)
```

The best result occurs for 27 PCs for which the RMSE is far better than anything achieved thus far. Unfortunately we have cheated, as we would not have access to the test sample in advance and so we would not know to use 27 components. We could reserve another part of our training dataset for model selection. We would now have two test sets — one for model selection and one for checking the performance. The downside of this is that we lose this additional test sample from our estimation which degrades its quality. Furthermore, there is the question of which and how many observations should go into this new test sample. We can avoid this dilemma with the use of *crossvalidation* (CV). We divide the data into m parts, equal or close in size. For each part, we use the rest of the data as the training set and that part as the test set. We evaluate the criterion of interest, RMSE in this case. We repeat for each part and average the result.

The `cross_val_predict` function from the `model_selection` module is helpful in performing the CV. By default, the data is randomly divided into ten parts for the CV. The random division means that the outcome is not deterministic. We have set the random number generator seed using `np.random.seed(123)` (and the choice of 123 is arbitrary) so that you will get the same result as here. You do not need to set the seed unless you want the results to be completely reproducible.

```
from sklearn.model_selection import cross_val_predict
rmsepcv = np.empty(len(ncomp))
np.random.seed(123)
for i in ncomp:
    pcsi = pcscores[:,:i]
    pcrmod.fit(pcsi,trainmeat.fat)
    ypred = cross_val_predict(pcrmod, pcsi, trainmeat.fat, cv=10)
    rmsepcv[i-1] = np.sqrt(mean_squared_error(
                   ypred,trainmeat.fat))
plt.plot(ncomp,rmsepcv)
plt.ylabel("RMSE")
plt.xlabel("Number of components")
```

Check where the minimum occurs:

```
np.argmin(rmsepcv)+1, min(rmsepcv)
(20, 2.6136)
```

The crossvalidated estimates of the RMSE are shown in the right panel of Figure 11.5. The minimum occurs at 20 components. We can find the RMSE on the test set from our previous calculation with

```
rmsep[np.argmin(rmsepcv)]
2.28674
```

This gives an RMSE on the test sample, 2.29, which is our best result so far.

11.2 Partial Least Squares

Partial least squares (PLS) is a method for relating a set of input variables $X_1,\ldots,X_m$ and outputs $Y_1,\ldots,Y_l$. PLS regression is comparable to PCR in that both predict the response using some number of linear combinations of the predictors. The difference is that while PCR ignores Y in determining the linear combinations, PLS regression explicitly chooses them to predict Y as well as possible.

We will consider only univariate PLS, i.e., $l = 1$ so that Y is scalar. We will look for models of the form:

$$\hat{y} = \beta_1 T_1 + \cdots + \beta_k T_k$$

where the T_i's are mutually orthogonal linear combinations of the Xs. See Figure 11.6.

Various algorithms have been presented for computing PLS. Most work by iteratively determining the T_is to predict y well, but at the same time maintaining orthogonality. One criticism of PLS is that it solves no well-defined modeling problem,

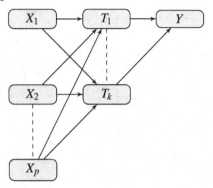

Figure 11.6   Schematic representation of partial least squares.

which makes it difficult to distinguish between the competing algorithms on theoretical rather than empirical grounds. Garthwaite (1994) presents an intuitive algorithm, but de Jong (1993) describes the SIMPLS method, which is among the best known. Several other algorithms exist.

Let's start by fitting a PLS model with only four components. We need the PLSRegression function from scikit-learn. We choose not to scale the predictors as they are already scaled.

```
from sklearn.cross_decomposition import PLSRegression
plsreg = PLSRegression(scale=False, n_components=4)
plsmod = plsreg.fit(Xtrain, trainmeat.fat)
```

As with principal components, we form linear combinations $T = XU$ of the predictors. These combinations are called the *loadings*. We have plotted the first three loadings in the left panel of Figure 11.7. This corresponds to Figure 11.3. PLS chooses somewhat different loadings from PCA because these are selected to better model the response.

```
plt.plot(frequency,plsmod.x_loadings_[:,0],'k-',label="1")
plt.plot(frequency,plsmod.x_loadings_[:,1],'k:',label="2")
plt.plot(frequency,plsmod.x_loadings_[:,2],'k--',label="3")
plt.legend()
plt.xlabel("Frequency")
plt.ylabel("Coefficient")
```

We can represent the fitted values as $\hat{y} = X\hat{\beta}$ so that $\hat{\beta}$ is the overall linear combination of the predictors used in this four component model as seen in the right panel of Figure 11.7.

```
plt.plot(frequency, plsmod.coef_)
plt.xlabel("Frequency")
plt.ylabel("Coefficient")
```

This corresponds to Figure 11.4 for the PCA fit. The combination is similar but not the same. We can check how well this model predicts the response in the training sample with:

```
np.sqrt(mean_squared_error(plsmod.predict(Xtrain),trainmeat.fat))
3.95
```

I have used the mean_squared_error function from the metrics module of

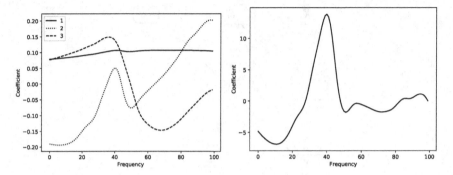

Figure 11.7   Linear combinations represented by the first three loadings of the PLSE model are shown on the left. The linear combination of the predictors used by the four-component PLS model is shown on the right.

scikit-learn. I prefer RMSE to MSE since it has units of the response. We see that the prediction performance is slightly better than the four-component PCA regression model. We expect this because PLS regression is designed to model the response.

We might do better with more than four components. We use crossvalidation on the training sample with ten folds:

```
ncomp = 50
rmsep = np.empty(ncomp)
component = np.arange(1, ncomp)
np.random.seed(123)
for i in range(ncomp):
    pls = PLSRegression(scale=False,n_components=i+1)
    ypred = cross_val_predict(pls, Xtrain, trainmeat.fat, cv=10)
    rmsep[i] = np.sqrt(mean_squared_error(ypred,trainmeat.fat))
plt.plot(range(1,ncomp+1),rmsep)
plt.ylabel("RMSE")
plt.xlabel("Number of components")
```

The plot of the RMSE as a function of the number of components is seen in Figure 11.8. We can find the minimizing number of components with:

```
np.argmin(rmsep)+1
14
```

We can implement this choice and check the performance on the training sample with:

```
plsbest = PLSRegression(scale=False,n_components=14)
plsbest.fit(Xtrain, trainmeat.fat)
ypred = plsbest.predict(Xtrain)
np.sqrt(mean_squared_error(ypred,trainmeat.fat))
1.95
```

But more importantly, the performance on the test sample is:

```
ytpred = plsbest.predict(Xtest)
np.sqrt(mean_squared_error(ytpred,testmeat.fat))
2.02
```

We have outperformed PCA regression while using fewer linear combinations.

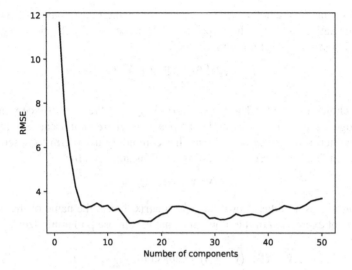

Figure 11.8   RMSE of PLS models computed using crossvalidation.

We have not checked any diagnostics in this analysis. PLS and PCR are just as sensitive to assumptions as OLS, so these are still mandatory in any full analysis.

PCR and PLS are particularly attractive methods when there are large numbers of predictors, p, relative to the sample size, n. They can still work even when $p > n$. This is not uncommon in some applications such as bioinformatics where data may give information on a large number of genes for a small sample of individuals.

PLS tends to have an advantage over PCR for prediction problems because PLS constructs its linear combination explicitly to predict the response. On the other hand, PCR is better suited for developing insights by forming linear combinations that have interesting interpretations.

Although both methods use the idea of dimension reduction, there is usually no reduction in the number of variables used since every predictor contributes to the linear combinations. If you are hoping to save the cost of measuring predictors in order to build a cheaper but still effective prediction model, the criterion-based variable selection as described in Chapter 10 or the lasso method described later in this chapter will be more useful.

11.3   Ridge Regression

Ridge regression makes the assumption that the regression coefficients (after normalization) should not be very large. This is a reasonable assumption in applications where you have a large number of predictors and you believe that many of them have some effect on the response. Hence shrinkage is embedded in the method. Ridge regression is particularly effective when the model matrix is collinear and the usual least squares estimates of β appear to be unstable.

Suppose that the predictors have been centered by their means and scaled by their standard deviations and that the response has been centered. The ridge regression estimate chooses the β that minimizes:

$$(y - X\beta)^T (y - X\beta) + \alpha \sum_j \beta_j^2$$

for some choice of $\alpha \geq 0$. The *penalty term* is $\sum_j \beta_j^2$. We want to keep this term small. Ridge regression is an example of a *penalized regression* because of the presence of this term. *Regularization* is another commonly-used term to describe this phenomenom. The ridge regression estimates of βs are given by:

$$\hat{\beta} = (X^T X + \alpha I)^{-1} X^T y$$

The αI introduces a "ridge" down the $X^T X$ matrix, hence the name of the method. An equivalent expression of the problem is that we choose β to minimize:

$$(y - X\beta)^T (y - X\beta) \quad \text{subject to} \quad \sum_{j=1}^{p} \beta_j^2 \leq t^2$$

where t takes the same role as α. We find the least squares solution subject to an upper bound on the size of the coefficients. The use of ridge regression can also be justified from a Bayesian perspective where a prior distribution on the parameters puts more weight on smaller values.

The nature of the ridge solution is seen in the left panel of Figure 11.9. The OLS fit that minimizes the residual sum of squares (RSS) is achieved at $\hat{\beta}_{LS}$. We can draw confidence ellipses around the least squares solution $\hat{\beta}_{LS}$ of increasing size. Points on these ellipses represent solutions for β which are progressively less desirable in terms of fit. We grow the ellipses until they intersect with a circle of radius t centered at the origin. This point of intersection will satisfy $\sum_{j=1}^{p} \beta_j^2 = t^2$ and be the best fitting solution subject to this requirement. Of course, if $\hat{\beta}_{LS}$ is already inside the circle, the ridge and OLS solutions are identical but usually t is small enough that this rarely occurs.

α (or t) may be chosen by automatic methods, but it is also sensible to plot the values of $\hat{\beta}$ as a function of α. You should pick the smallest value of α that produces stable estimates of β.

We demonstrate the method on the meat spectroscopy data; $\alpha = 0$ corresponds to least squares while we find that as $\alpha \to \infty$: $\hat{\beta} \to 0$. In practice, we are interested in a narrower range of α. Here we set up a log-spaced grid of α values between 10^{-5} and 10^{-10}. For other datasets, you may need to vary this range to capture the interesting region.

```
n_alphas = 50
alphas = np.logspace(-10, -5, n_alphas)
```

For each value of α on the grid, we fit the ridge regression and save the coefficients.

```
coefs = []
for a in alphas:
    ridge = linear_model.Ridge(alpha=a)
    ridge.fit(Xtrain, trainmeat.fat)
    coefs.append(ridge.coef_)
```

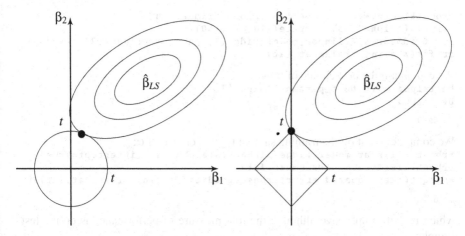

Figure 11.9 Ridge and lasso regression are illustrated. On the left, confidence ellipses of increasing level are plotted around the least squares estimate. The largest ellipse intersects the circle of radius t at the ridge estimate. On the right, the largest ellipse intersects the square at the lasso estimate.

Now we plot the coefficients as a function of α as seen in the left panel of Figure 11.10. This is called the *ridge trace plot*.

```
ax = plt.gca()
ax.plot(alphas, coefs, 'k', alpha=0.2)
ax.set_xscale('log')
ax.set_xlim(ax.get_xlim()[::-1])
plt.xlabel('alpha')
plt.ylabel('Coefficients')
```

The *ridge trace plot* is shown in Figure 11.10.

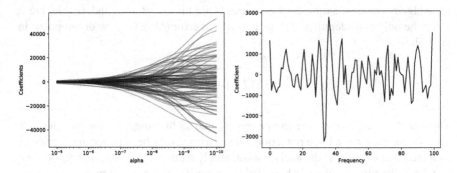

Figure 11.10 The ridge trace plot for the meat spectroscopy data is shown on the left. On the right, we have the coefficients for the optimal amount of shrinkage.

We can select the value of α using crossvalidation by searching across a grid of α values:

```
from sklearn.model_selection import GridSearchCV
pars = {'alpha':np.logspace(-10, -5, 50)}
rr= GridSearchCV(linear_model.Ridge(), pars, scoring='r2', cv=10)
rr.fit(Xtrain, trainmeat.fat)
```

We can obtain the best choice of α as:
```
bestalpha = rr.best_params_['alpha']
bestalpha
```
9.54e-07

We compute the fit on the training sample for this choice of α:
```
rrbest = linear_model.Ridge(alpha=bestalpha, fit_intercept=True)
rrbest.fit(Xtrain, trainmeat.fat)
np.sqrt(mean_squared_error(rrbest.predict(Xtrain),trainmeat.fat))
```
1.56

which is quite impressive although the true measure of performance is on the test sample:
```
np.sqrt(mean_squared_error(rrbest.predict(Xtest),testmeat.fat))
```
2.03

We have surpassed the PLS result.

It is interesting to look at the regression coefficients for this choice of α as seen in the right panel of Figure 11.10.
```
plt.plot(frequency,rrbest.coef_)
plt.xlabel("Frequency")
plt.ylabel("Coefficient")
```

These should be compared to the PCA regression coefficients seen in Figure 11.4. Compared to the four-component PCA, the ridge coefficients are not very smooth. But we have found that having four components is insufficient for good prediction. The best amount of shrinkage is not as extreme. Compared to the full linear model, the ridge coefficients are very much smaller.

Ridge regression estimates of coefficients are biased. Bias is undesirable, but it is not the only consideration. The mean-squared error (MSE) can be decomposed in the following way:

$$E(\hat{\beta} - \beta)^2 = (E(\hat{\beta} - \beta))^2 + E(\hat{\beta} - E\hat{\beta})^2$$

Thus the MSE of an estimate can be represented as the square of the bias plus the variance. Sometimes a large reduction in the variance may be obtained at the price of an increase in the bias. If the MSE is substantially reduced as a consequence, then we may be willing to accept some bias. This is the trade-off that ridge regression makes — a reduction in variance at the price of an increase in bias. This is a common dilemma.

Frank and Friedman (1993) compared PCR, PLS and ridge regression and found the best results for ridge regression. Of course, for any given dataset any of the methods may prove to be the best, so picking a winner is difficult.

11.4   Lasso

The lasso method is apparently very similar to the ridge regression method. We choose $\hat{\beta}$ to minimize:

$$(y - X\beta)^T (y - X\beta) \quad \text{subject to} \quad \sum_{j=1}^{p} |\beta_j| \leq t$$

or equivalently to minimize:

$$(y - X\beta)^T (y - X\beta) + \alpha \sum_j \beta_j^2$$

for some choice of $\alpha \geq 0$. The penalty term differs in form from the ridge case. This method was introduced by Tibshirani (1996). There is no explicit solution to this problem as in the ridge regression case although it can be found quite efficiently as described within the context of a more general method called *least angle regression* as given in Efron, Hastie, Johnstone, and Tibshirani (2004).

The important difference between lasso and ridge regression is in the nature of the solutions which is illustrated in Figure 11.9. For the lasso, the L_1 constraint of $\sum_{j=1}^{p} |\beta_j| \leq t$ defines a square in two dimensions as seen in the figure. In higher dimensions, it defines a polytope with vertices on the coordinates' axes and edges where some number of the coordinate values will be zero. In the figure, we can see that the expanding ellipses touch the square at a vertex. We can see that, depending on the value of $\hat{\beta}_{LS}$, the correlation of variables which determines the orientation of the ellipse and the value of t, that a solution on a vertex will occur frequently. In higher dimensions, the solution will also commonly lie on an edge or vertex of the polytope. As t increases, more variables are added to the model and their coefficients become larger. For large enough t, the restriction of $\sum_{j=1}^{p} |\beta_j| \leq t$ is redundant and the least squares solution is returned.

For the lasso, for moderate values of t, many $\hat{\beta}_j$ tend to be zero. The use of lasso is most appropriate when we believe the effects are *sparse* — that the response can be explained by a small number of predictors with the rest having no effect. This means that lasso can be regarded as a type of variable selection method because when $\hat{\beta}_j = 0$, the corresponding predictor x_j is effectively eliminated from the regression. In contrast, ridge regression does not eliminate any variables; it only makes the $\hat{\beta}_j$ smaller.

We return to the fat density data used earlier in the chapter and use only the circumference measures as predictors.

```
import faraway.datasets.fat
fat = faraway.datasets.fat.load()
X = fat.iloc[:,8:]
```

As with the ridge regression, we must choose a grid of possible values for α:

```
n_alphas = 50
alphas = np.logspace(-2, 2, n_alphas)
```

You may need to change the range for other datasets. We now fit the lasso model for each of these αs and save the coefficients:

```
lasso = linear_model.Lasso()
coefs = []
for a in alphas:
    lasso.set_params(alpha=a)
    lasso.fit(X, fat.brozek)
    coefs.append(lasso.coef_)
```

As before, we can use crossvalidation to select the best value of α:

```
lassocv = linear_model.LassoCV(cv = 10)
lassocv.fit(X, fat.brozek)
lassocv.alpha_
0.096017
```

The LassoCV selects a grid of α values. It is wise to check the range of these values:

```
min(lassocv.alphas_),max(lassocv.alphas_)
(0.0677, 67.7)
```

We see that the optimized value of 0.096 falls in the interior of this range giving us some confidence that we have found a global optimum. If we land on the edge of the range, we would need to supply a shifted grid of αs to continue the search for the optimum.

We plot the coefficients for each value of α in Figure 11.11:

```
ax = plt.gca()
ax.plot(alphas, coefs)
plt.xlim(min(alphas)/5,max(alphas))
ax.set_xscale('log')
for i in range(len(X.columns)):
    plt.text(min(alphas)/4,coefs[0][i],X.columns[i])
plt.xlabel('alpha')
plt.ylabel('coefficients')
plt.axvline(lassocv.alpha_)
```

For each predictor, the estimated coefficient will decrease as the value of α increases. These paths for each predictor are shown in the plot. The crossvalidated choice of α is shown as a vertical line. For larger values of α, only one predictor, abdomen circumference enters into the model. As α is reduced, other predictors are added to the model: first hip, then neck, then others. At the smallest values of α, all the predictors are in the model and the coefficients tend to their least squares values. Thus the choice of α can be used as a variable selection method.

In this example, crossvalidation chooses a relatively small value of α which includes most predictors in the model. We can examine these:

```
lassocv.coef_.round(3)
array([-0.553, -0.078,  0.938, -0.373,  0.117, -0.062, -0.   ,  0.076,
        0.277, -1.144])
```

Only one the estimates is set to zero — the ankle circumference. By comparison, we might look at the least squares estimates:

```
reg = linear_model.LinearRegression(fit_intercept=True)
reg.fit(X, fat.brozek)
reg.coef_.round(3)
array([-0.582, -0.091,  0.96 , -0.391,  0.134, -0.094,  0.004,  0.111,
        0.345, -1.353])
```

Lasso is more useful for problems with much larger numbers of variables such as the ongoing spectroscopy example of this chapter. This example is problematic for the lasso due to the strong collinearity in the predictors which causes convergences difficulties with the estimator. Using crossvalidation, we find that a very small value

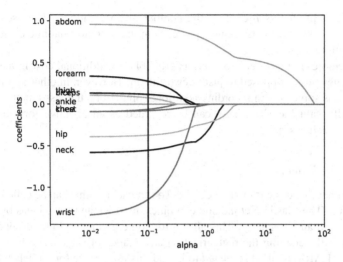

Figure 11.11   Lasso plots for the fat density data. The coefficients as a function of α are shown. The crossvalidated choice of α is shown as a vertical line.

of α is indicated. We also set a large number of iterations and a relaxed convergence criterion to help with the computation:

```
lasso = linear_model.Lasso(max_iter=1e6,tol=0.001)
lasso.set_params(alpha=1e-9)
lasso.fit(Xtrain, trainmeat.fat)
```

The small value of α indicates very little shrinkage. We check the proportion of nonzero coefficients in this fit:

```
np.mean(abs(lasso.coef_)>1e-5)
1.0
```

None of the lasso coefficients are zero. We check the performance on the training sample:

```
np.sqrt(mean_squared_error(lasso.predict(Xtrain),trainmeat.fat))
2.39
```

and on the test sample:

```
np.sqrt(mean_squared_error(lasso.predict(Xtest),testmeat.fat))
2.35
```

The results are worse than PCR, PLS and ridge regression although still much better than the full least squares model. We could address the convergence problems by scaling the data although this would require some additional effort in scaling the test sample in the same way and making the results comparable to the other methods. Even so, this data example is not well suited for the lasso. We have strongly collinear predictors and we have evidence from the previous analyses that most of them are helpful in predicting the response. Lasso excels in datasets where we expect only a few of the predictors to have an effect on the response. Lasso works well in these *sparse* effects situations. For example, this arises in the analysis of gene expression data where we have good reason to believe that only a small number of genes have an influence on the quantity of interest. Lasso is also useful in that it still works when the number of predictors p exceeds the number of observations. The use of lasso in applications where effects are not sparse, as in many socioeconomic examples, is less compelling.

Furthermore, in this example, there is probably no additional cost in measuring all the frequencies as opposed to just a few of them. However, in other applications, where there is some cost in recording additional predictors, the lasso method would be especially valuable. The lasso can be regarded as an effective form of model selection in such cases.

11.5   Other Methods

There are other choices within the `scikit-learn` package that can be called *shrinkage* methods. The Elastic-Net method combines the ridge and lasso ideas by having both an L_1 and an L_2 penalty. This allows for some predictors to be dropped as in lasso while still retaining the regularization advantages of ridge. The Least Angle Regression (LARS) method is related to lasso in its preference for models with a reduced number of predictors. The Orthogonal Matching Pursuit method goes further than lasso in just encouraging the elimination of predictors — it specifies a maximum number of nonzero coefficients. The `scikit-learn` package also contains a Bayesian regression implementation. By imposing weakly informative priors on the parameters we achieve a similar effect to ridge regression. It is possible to make an exact identification between the two methods.

There are many other methods for predicting a continuous response as a function of some predictors. Neural networks, support vector machines, random forests and more can all be used for regression problems. But these methods do not use the linear form for the combination of the predictors. Although these methods can be very effective in prediction, they do not have the interpretable parameter of the model coefficients which is a key advantage of the methods described in this chapter. For all four methods, we can see clearly how new predictions will be generated as more data is collected because of the transparent linear form.

Exercises

1. Use the `seatpos` data with `hipcenter` as the response.
   (a) Use the predictors: `HtShoes`, `Ht`, `Seated`, `Arm`, `Thigh` and `Leg`. Perform a PCA on these unscaled predictors and compute the standard deviations of the principal components. How many components should be chosen?
   (b) What are the loadings on the first principal component? Give an interpretation.
   (c) Fit a linear model with all six predictors. Report on the R-squared. Which predictors are statistically significant?
   (d) Fit a linear model with all six principal components. How does the overall fit differ from the linear model of the previous question? Which principal components are statistically significant?
   (e) Fit a linear model with only the height as a predictor. Make an F-test comparing

it to the model of the previous question. Explain why the F-test is valid here. Should this model be preferred?

(f) Now include Age and Weight in the set of predictors. Repeat the PCA but first scale the predictors. Explain why scaling is reasonable. How many components should be chosen?

(g) Produce the loadings on the first and second PCs. Interpret.

(h) Fit a two-predictor linear model inspired by the last PCA analysis. Comment.

2. Use the seatpos data with all variables (scaled) as predictors except hipcenter. Use the robust PCA method to identify any unusual cases. Say what is unusual about them.

3. Fit PLS models to the seatpos data with all variables as predictors but with hipcenter as the response.

(a) Use unscaled predictors. Select an appropriate number of components.

(b) Now scale the predictors to standard units. Select an appropriate number of components.

(c) Which version should be preferred - scaled or unscaled?

4. Fit a ridge regression model to the seatpos data with hipcenter as the response and all other variables as (scaled) predictors.

(a) Produce the ridge trace plot with each curve labeled with the predictor name. You will need to choose the range of alpha carefully.

(b) Use crossvalidation to choose the optimum amount of shrinkage.

(c) Extract the estimated coefficients at the crossvalidated choice of alpha. Interpret.

5. Fit a LASSO regression model to the seatpos data with hipcenter as the response and all other variables as (scaled) predictors.

(a) Determine the optimal value of the smoothing parameter using crossvalidation.

(b) Make a plot of the estimated coefficients as a function of alpha. Label the coefficient paths. Show the optimal choice of alpha. Is there anything unusual about the plot?

(c) Extract and interpret the coefficients at the optimum value of alpha.

Chapter 12

Insurance Redlining — A Complete Example

In this chapter, we present a relatively complete data analysis. The example is interesting because it illustrates several of the ambiguities and difficulties encountered in statistical practice.

Insurance redlining refers to the practice of refusing to issue insurance to certain types of people or within some geographic area. The name comes from the act of drawing a red line around an area on a map. Now few would quibble with an insurance company refusing to sell auto insurance to a frequent drunk driver, but other forms of discrimination would be unacceptable.

In the late 1970s, the US Commission on Civil Rights examined charges by several Chicago community organizations that insurance companies were redlining their neighborhoods. Because comprehensive information about individuals being refused homeowners insurance was not available, the number of FAIR plan policies written and renewed in Chicago by zip code for the months of December 1977 through May 1978 was recorded. The FAIR plan was offered by the city of Chicago as a default policy to homeowners who had been rejected by the voluntary market. Information on other variables that might affect insurance writing such as fire and theft rates was also collected at the zip code level. The variables are:

race racial composition in percentage of minority

fire fires per 100 housing units

theft thefts per 1000 population

age percentage of housing units built before 1939

involact new FAIR plan policies and renewals per 100 housing units

income median family income in thousands of dollars

side north or south side of Chicago

The data come from Andrews and Herzberg (1985) where more details of the variables and the background are provided.

12.1 Ecological Correlation

Notice that we do not know the races of those denied insurance. We only know the racial composition in the corresponding zip code. This is an important difficulty that needs to be considered before starting the analysis.

When data are collected at the group level, we may observe a correlation between two variables. The ecological fallacy is concluding that the same correlation holds

197

at the individual level. For example, in countries with higher fat intakes in the diet, higher rates of breast cancer have been observed. Does this imply that individuals with high fat intakes are at a higher risk of breast cancer? Not necessarily. Relationships seen in observational data are subject to confounding, but even if this is allowed for, bias is caused by aggregating data. First we load the packages:

```
import numpy as np
import matplotlib.pyplot as plt
import statsmodels.api as sm
import statsmodels.formula.api as smf
import pandas as pd
import faraway.utils
```

We consider an example taken from US demographic data:

```
import faraway.datasets.eco
eco = faraway.datasets.eco.load()
plt.scatter(eco.usborn, eco.income)
plt.ylabel("Income")
plt.xlabel("Fraction US born")
```

In the left panel of Figure 12.1, we see the relationship between 1998 per capita income dollars from all sources and the proportion of legal state residents born in the United States in 1990 for each of the 50 states plus the District of Columbia (D.C.). We can see a clear negative correlation. We can fit a regression line and show the

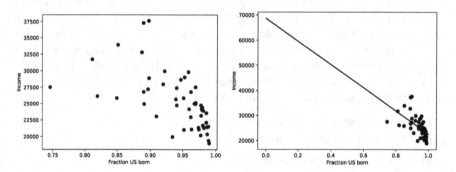

Figure 12.1   1998 annual per capita income and proportion US born for 50 states plus D.C. The plot on the right shows the same data as on the left, but with an extended scale and the least squares fit shown.

fitted line on an extended range:

```
lmod = sm.OLS(eco.income, sm.add_constant(eco.usborn)).fit()
lmod.sumary()
```

```
           coefs     stderr tvalues pvalues
const   68,642.239 8,739.004    7.85  0.0000
usborn -46,018.639 9,279.116   -4.96  0.0000
```

```
n=51 p=2 Residual SD=3489.541 R-squared=0.33
```

Now show the regression line on the data as seen in the right panel of Figure 12.1:

```
plt.scatter(eco.usborn, eco.income)
plt.ylabel("Income")
plt.xlabel("Fraction US born")
```

```
xr = np.array([0,1])
plt.plot(xr,lmod.params[0] + xr*lmod.params[1])
```
We see that there is a clear statistically significant relationship between the per capita annual income and the proportion who are US born. What does this say about the average annual income of people who are US born and those who are naturalized citizens? If we substitute `usborn=1` into the regression equation, we get 68642 − 46019 = \$22,623, while if we put `usborn=0`, we get \$68,642. This suggests that on average, naturalized citizens earn three times more than US born citizens. In truth, information from the US Bureau of the Census indicates that US born citizens have an average income just slightly larger than naturalized citizens. What went wrong with our analysis?

The ecological inference from the aggregate data to the individuals requires an assumption of constancy. Explicitly, the assumption would be that the incomes of the native born do not depend on the proportion of native born within the state (and similarly for naturalized citizens). This assumption is unreasonable for these data because immigrants are naturally attracted to wealthier states.

This assumption is also relevant to the analysis of the Chicago insurance data since we have only aggregate data. We must keep in mind that the results for the aggregated data may not hold true at the individual level.

12.2 Initial Data Analysis

Start by reading the data in and examining it:
```
import faraway.datasets.chredlin
chredlin = faraway.datasets.chredlin.load()
chredlin.head()
```
```
    zip  race  fire  theft  age  involact  income side
0  60626  10.0   6.2    29  60.4      0.0  11.744    n
1  60640  22.2   9.5    44  76.5      0.1   9.323    n
2  60613  19.6  10.5    36  73.5      1.2   9.948    n
3  60657  17.3   7.7    37  66.9      0.5  10.656    n
4  60614  24.5   8.6    53  81.4      0.7   9.730    n
```
Summarize:
```
chredlin.drop('zip',1).describe().round(2)
```
```
        race   fire   theft    age  involact  income
count  47.00  47.00   47.00  47.00     47.00   47.00
mean   34.99  12.28   32.36  60.33      0.61   10.70
std    32.59   9.30   22.29  22.57      0.63    2.75
min     1.00   2.00    3.00   2.00      0.00    5.58
25%     3.75   5.65   22.00  48.60      0.00    8.45
50%    24.50  10.40   29.00  65.00      0.40   10.69
75%    57.65  16.05   38.00  77.30      0.90   11.99
max    99.70  39.70  147.00  90.10      2.20   21.48
```
and for the categorical predictor, `side`:
```
chredlin.side.value_counts()
```
```
n   25
s   22
```
We see that there is a wide range in the `race` variable, with some zip codes almost entirely minority or non-minority. This is good for our analysis since it will reduce the variation in the regression coefficient for race, allowing us to assess this effect

more accurately. If all the zip codes were homogeneous, we would never be able to discover an effect from these aggregated data. We also note some skewness in the theft and income variables. The response involact has a large number of zeroes. This is not good for the assumptions of the linear model, but we have little choice but to proceed. We will not use the information about north versus south side until later. Now make some graphical summaries:

```python
fig, ((ax1, ax2, ax3), (ax4, ax5, ax6)) = \
    plt.subplots(nrows=2, ncols=3, sharey=True)
ax1.scatter(chredlin.race, chredlin.involact)
ax1.set(title="Race")
ax1.set_ylabel("Involact")

ax2.scatter(chredlin.fire, chredlin.involact)
ax2.set(title="Fire")

ax3.scatter(chredlin.theft, chredlin.involact)
ax3.set(title="Theft")

ax4.scatter(chredlin.age, chredlin.involact)
ax4.set(title="Age")
ax4.set_ylabel("Involact")

ax5.scatter(chredlin.income, chredlin.involact)
ax5.set(title="Income")

ax6.scatter(chredlin.side, chredlin.involact)
ax6.set(title="Side")

fig.tight_layout()
```

The plots are seen in Figure 12.2. We can see some outlier and influential points. We can also see that the fitted line sometimes goes below zero which is problematic since observed values of the response cannot be negative. Let's focus on the relationship between involact and race:

```python
lmod = sm.OLS(chredlin.involact,
    sm.add_constant(chredlin.race)).fit()
lmod.sumary()
```

	coefs	stderr	tvalues	pvalues
const	0.129	0.097	1.34	0.1878
race	0.014	0.002	6.84	0.0000

```
n=47 p=2 Residual SD=0.449 R-squared=0.51
```

We can clearly see that homeowners in zip codes with a high percentage of minorities are taking the default FAIR plan insurance at a higher rate than other zip codes. That is not in doubt. However, can the insurance companies claim that the discrepancy is due to greater risks in some zip codes? The insurance companies could claim that they were denying insurance in neighborhoods where they had sustained large fire-related losses and any discriminatory effect was a by-product of legitimate business practice. We plot some of the variables involved by this question in Figure 12.3.

```python
fig = plt.figure(figsize=(6,3))

ax1 = fig.add_subplot(121)
```

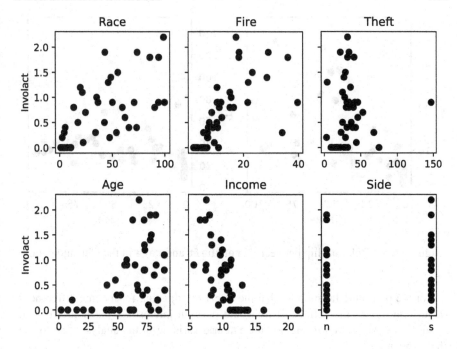

Figure 12.2   Plots of the Chicago insurance data.

```
ax1.scatter(chredlin.race, chredlin.fire)
ax1.set_xlabel("Race")
ax1.set_ylabel("Fire")

ax2 = fig.add_subplot(122)
ax2.scatter(chredlin.race, chredlin.theft)
ax2.set_xlabel("Race")
ax2.set_ylabel("Theft")

fig.tight_layout()
```

We can see that there is indeed a relationship between the fire rate and the percentage of minorities. We also see that there is large outlier that may have a disproportionate effect on the relationship between the theft rate and the percentage of minorities.

The question of which variables should also be included in the regression so that their effect may be adjusted for is difficult. Statistically, we can do it, but the important question is whether it should be done at all. For example, it is known that the incomes of women in the United States and other countries are generally lower than those of men. However, if one adjusts for various predictors such as type of job and length of service, this gender difference is reduced or can even disappear. The controversy is not statistical but political — should these predictors be used to make the adjustment?

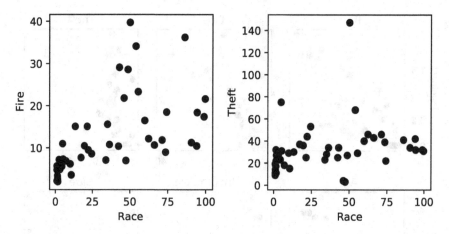

Figure 12.3   Relationship between fire, theft and race in the Chicago data.

For the present data, suppose that the effect of adjusting for income differences was to remove the race effect. This would pose an interesting, but nonstatistical, question. I have chosen to include the income variable in the analysis just to see what happens.

I have decided to use log(income) partly because of skewness in this variable, but also because income is better considered on a multiplicative rather than additive scale. In other words, $1,000 is worth a lot more to a poor person than to a millionaire because $1,000 is a much greater fraction of the poor person's wealth.

12.3   Full Model and Diagnostics

We start with the full model:

```
lmod = smf.ols(
    'involact ~ race + fire + theft + age + np.log(income)',
    chredlin).fit()
lmod.sumary()
```

	coefs	stderr	tvalues	pvalues
Intercept	-1.186	1.100	-1.08	0.2876
race	0.010	0.002	3.82	0.0004
fire	0.040	0.009	4.55	0.0000
theft	-0.010	0.003	-3.65	0.0007
age	0.008	0.003	3.04	0.0041
np.log(income)	0.346	0.400	0.86	0.3925

n=47 p=6 Residual SD=0.335 R-squared=0.75

Before leaping to any conclusions, we should check the model assumptions. These two diagnostic plots are seen in Figure 12.4:

```
plt.scatter(lmod.fittedvalues, lmod.resid)
plt.ylabel("Residuals")
plt.xlabel("Fitted values")
plt.axhline(0)
```

```
sm.qqplot(lmod.resid, line="q")
```

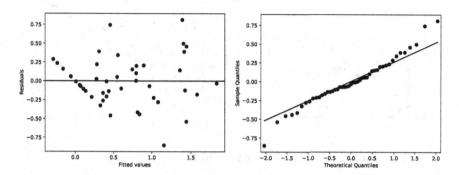

Figure 12.4   Diagnostic plots of the initial model for the Chicago insurance data.

The diagonal streak in the residual-fitted plot is caused by the large number of zero response values in the data. When $y = 0$, the residual $\hat{\varepsilon} = -\hat{y} = -x^T\hat{\beta}$, hence the line. Turning a blind eye to this feature, we see no particular problem. The Q–Q plot looks fine too. This is reassuring since we know from the form of the response with so many zero values that it cannot possibly be normally distributed. We'll rely on the central limit theorem, the size of the sample and lack of long-tailed or skewed residuals to be comfortable with the reported p-values.

We now look for transformations. We try some partial residual plots as seen in Figure 12.5. First for race:

```
pr = lmod.resid + chredlin.race*lmod.params['race']
plt.scatter(chredlin.race, pr)
plt.xlabel("Race")
plt.ylabel("partial residuals")
xr = np.array(plt.xlim())
plt.plot(xr, xr*lmod.params['race'])
```

and for fire:

```
pr = lmod.resid + chredlin.fire*lmod.params['fire']
plt.scatter(chredlin.fire, pr)
plt.xlabel("Fire")
plt.ylabel("partial residuals")
xr = np.array(plt.xlim())
plt.plot(xr, xr*lmod.params['fire'])
```

These plots indicate no need to transform. It would have been inconvenient to transform the race variable since that would have made interpretation more difficult. Fortunately, we do not need to worry about this. We examined the other partial residual plots and experimented with polynomials for the predictors. No transformation of the predictors appears to be worthwhile.

We choose to avoid a transformation of the response. The zeroes in the response would have restricted the possibilities and furthermore would have made interpretation more difficult. A square root transformation is possible but whatever slim advantage this might offer, it makes explanation more problematic.

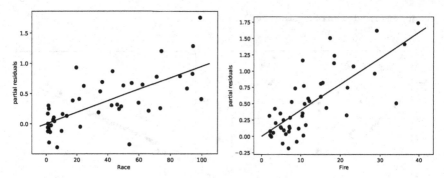

Figure 12.5   Partial residual plots for `race` and `fire`.

12.4   Sensitivity Analysis

How robust is our conclusion to the choice of covariates used to adjust the response? In the full model used earlier, we used all four covariates but we may wonder how sensitive our findings are to this choice. Certainly, one might question whether we should adjust the response for the average income of the zip code. Other objections or uncertainties might be raised by use of the other covariates also.

We can investigate these concerns by fitting other models that vary the choice of adjusting covariates. In this example, there are four such covariates and so there are only 16 possible combinations in which they may be added to the model. It is practical to fit and examine all these models.

The mechanism for creating all 16 models is rather complex and you may wish to skip to the output. First we create all subsets of $(1,2,3,4)$.

```
import itertools
inds = [1, 2, 3, 4]
clist = []
for i in range(0, len(inds)+1):
    clist.extend(itertools.combinations(inds, i))
```

Now we fit all 16 possible models:

```
X = chredlin.iloc[:,[0,1,2,3,5]].copy()
X.loc[:,'income'] = np.log(chredlin['income'])
betarace = []
pvals = []
for k in range(0, len(clist)):
    lmod = sm.OLS(chredlin.involact,
        sm.add_constant(X.iloc[:,np.append(0,clist[k])])).fit()
    betarace.append(lmod.params[1])
    pvals.append(lmod.pvalues[1])
```

Construct the variable names

```
vlist = ['race']
varnames = np.array(['race','fire','theft','age','logincome'])
for k in range(1, len(clist)):
    vlist.append('+'.join(varnames[np.append(0,clist[k])]))
```

and create the output:

```
pd.DataFrame({'beta':np.round(betarace,4),
    'pvals':np.round(pvals,4)}, index=vlist)
```

	beta	pvals
race	0.0139	0.0000
race+fire	0.0089	0.0002
race+theft	0.0141	0.0000
race+age	0.0123	0.0000
race+logincome	0.0082	0.0087
race+fire+theft	0.0082	0.0002
race+fire+age	0.0089	0.0001
race+fire+logincome	0.0070	0.0160
race+theft+age	0.0128	0.0000
race+theft+logincome	0.0084	0.0083
race+age+logincome	0.0099	0.0017
race+fire+theft+age	0.0081	0.0001
race+fire+theft+logincome	0.0073	0.0078
race+fire+age+logincome	0.0085	0.0041
race+theft+age+logincome	0.0106	0.0010
race+fire+theft+age+logincome	0.0095	0.0004

The output shows the $\hat{\beta}_1$ and the associated *p*-values for all 16 models. We can see see that the value of $\hat{\beta}_1$ varies somewhat with a high value about double the low value. But in no case does the *p*-value rise above 5%. So although we may have some uncertainty over the magnitude of the effect, we can be sure that the significance of the effect is not sensitive to the choice of adjusters.

Suppose the outcome had not been so clear-cut and we were able to find models where the predictor of interest (in this case, race) was not statistically significant. The investigation would then have become more complex, because we would need to consider more deeply which covariates should be adjusted for and which should not. Such a discussion is beyond the scope of this book but illustrates why causal inference is a difficult subject.

We should also be concerned whether our conclusions are sensitive to the inclusion or exclusion of a small number of cases. Influence diagnostics are useful for this purpose. We start by computing the scaled change in the coefficients caused by omitting a single case. In particular, we are interested if the omission of a single case could cause race variable to become statistically significant:

```
lmod = smf.ols(
    'involact ~ race + fire + theft + age + np.log(income)',
    chredlin).fit()
diagv = lmod.get_influence()
min(diagv.dfbetas[:,1])
-0.39954
```

We see that the largest reduction in the t-statistic is about 0.4. Since the full data value is 3.82, this reduction is insufficient to cause a change to our conclusions.

It is also worth considering the influence on the adjustment covariates. We plot the leave-out-one differences in the scaled $\hat{\beta}$ for theft and fire as seen in the left panel of Figure 12.6:

```
plt.scatter(diagv.dfbetas[:,2], diagv.dfbetas[:,3])
plt.xlabel("Change in Fire")
plt.ylabel("Change in Theft")
plt.axhline(0)
plt.axvline(0)
```

Let's also take a look at the standardized residuals and leverage which can be conveniently constructed with:

```
sm.graphics.influence_plot(lmod)
```
The diameter of the circles correspond to the size of the Cook Statistic. We see that
cases 5 and 23, as seen in the right panel of Figure 12.6, stick out.

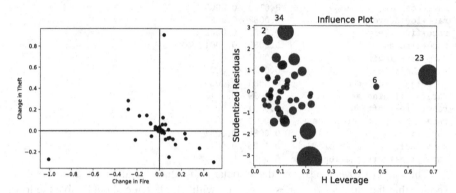

Figure 12.6   Plot of the leave-out-one coefficient differences is shown on the left.
Plot of the standardized residuals against the leverages is shown on the right.

It is worth looking at other leave-out-one coefficient plots also. We also notice
that there is no standardized residual extreme enough to call an outlier. Let's take a
look at the two cases:
```
chredlin.iloc[[5, 23],:]
       zip  race  fire  theft   age  involact  income side
5    60610  54.0  34.1     68  52.6       0.3   8.231    n
23   60607  50.2  39.7    147  83.0       0.9   7.459    n
```
These are high theft and fire zip codes. See what happens when we exclude these
points in the model:
```
ch45 = chredlin.drop(chredlin.index[[5,23]])
lmod45 = smf.ols(
    'involact ~ race + fire + theft + age + np.log(income)',
    ch45).fit()
lmod45.sumary()
                coefs stderr tvalues pvalues
Intercept      -0.577  1.080   -0.53  0.5964
race            0.007  0.003    2.62  0.0126
fire            0.050  0.009    5.79  0.0000
theft          -0.006  0.004   -1.48  0.1471
age             0.005  0.003    1.79  0.0818
np.log(income)  0.116  0.401    0.29  0.7745
```

n=45 p=6 Residual SD=0.303 R-squared=0.80

The coefficient for race is reduced compared to the full data fit but remains
statistically significant. Furthermore, theft and age are no longer significant.

So we have verified that our conclusions are also robust to the exclusion of one
or perhaps two cases from the data. This is reassuring since a conclusion based on
the accuracy of measurement for a single case would be a cause for concern. If this
problem did occur, we would need to be particularly sure of these measurements. In
some situations, one might wish to drop such influential cases, but this would require

strong arguments that such points were in some way exceptional. In any case, it would be very important to disclose this choice in the analysis.

Now if we try very hard to poke a hole in our result, we can find this model where two cases have been dropped:

```
lmodalt = smf.ols('involact ~ race + fire + np.log(income)',
    ch45).fit()
lmodalt.sumary()
```

	coefs	stderr	tvalues	pvalues
Intercept	0.753	0.836	0.90	0.3728
race	0.004	0.002	1.85	0.0718
fire	0.051	0.008	6.04	0.0000
np.log(income)	-0.362	0.319	-1.14	0.2628

n=45 p=4 Residual SD=0.309 R-squared=0.79

In this model, `race` no longer meets the threshold for significance. However, there is no compelling reason to advocate for this model against the large weight of other alternatives we have considered.

This illustrates a wider problem with regression modeling in that the data usually do not unequivocally suggest one particular model. It is easy for independent analysts to apply similar methods but in different orders and in somewhat different ways resulting in different model choices. See Faraway (1994) for some examples. For this reason, the good analyst explores the data thoroughly and considers multiple models. One might settle on one final model, but confidence in the conclusions will be enhanced if it can be shown that competing models result in similar conclusions. Our analysis in this chapter demonstrates this concern for alternatives, but there is an unavoidable reliance on human judgment. An unscrupulous analyst can explore a large number of models but report only the one that favors a particular conclusion.

A related concept is *model uncertainty*. We surely do not know the true model for this data and somehow our conclusions should reflect this. The regression summary outputs provide standard errors and *p*-values that express our uncertainty about the parameters of the model, but they do not reflect the uncertainty about the model itself. This means that we will tend to be more confident about our inferences than is justified. There are several possible ways to mitigate this problem. One simple approach is data splitting as used in the running example on the `meatspec` data in Chapter 11. Another idea is to bootstrap the whole data analysis as demonstrated by Faraway (1992). Alternatively, it may be possible to use *model averaging* as in Raftery, Madigan, and Hoeting (1997).

12.5 Discussion

There is some ambiguity in the conclusion here. These reservations have several sources. There is some doubt because the response is not a perfect measure of people being denied insurance. It is an aggregate measure that raises the problem of ecological correlations. We have implicitly assumed that the probability a minority homeowner would obtain a FAIR plan after adjusting for the effect of the other covariates is constant across zip codes. This is unlikely to be true. If the truth is simply a variation about some constant, then our conclusions will still be reasonable, but if this probability varies in a systematic way, then our conclusions may be off the mark. It would be a very good idea to obtain some individual level data.

We have demonstrated statistical significance for the effect of race on the response. But statistical significance is not the same as practical significance. The largest value of the response is only 2.2% and most other values are much smaller. Using our preferred models, the predicted difference between 0% minority and 100% minority is about 1%. So while we may be confident that some people are affected, there may not be so many of them. We would need to know more about predictors like insurance renewal rates to say much more but the general point is that the size of the p-value does not tell you much about the practical size of the effect.

There is also the problem of a potential latent variable that might be the true cause of the observed relationship. Someone with first-hand knowledge of the insurance business might propose one. This possibility always casts a shadow of doubt on our conclusions.

Another issue that arises in cases of this nature is how much the data should be aggregated. For example, suppose we fit separate models to the two halves of the city. Fit the model to the south of Chicago:

```
lmods = smf.ols('involact ~ race + fire + theft + age',
     chredlin.loc[chredlin.side == 's',:]).fit()
lmods.sumary()
```

```
          coefs stderr tvalues pvalues
Intercept -0.234  0.238   -0.99  0.3380
race       0.006  0.003    1.81  0.0873
fire       0.048  0.017    2.87  0.0107
theft     -0.007  0.008   -0.79  0.4423
age        0.005  0.005    0.99  0.3348
```

```
n=22 p=5 Residual SD=0.351 R-squared=0.74
```

and now to the north:

```
lmodn = smf.ols('involact ~ race + fire + theft + age',
     chredlin.loc[chredlin.side == 'n',:]).fit()
lmodn.sumary()
```

```
          coefs stderr tvalues pvalues
Intercept -0.319  0.227   -1.40  0.1759
race       0.013  0.004    2.81  0.0109
fire       0.023  0.014    1.65  0.1135
theft     -0.008  0.004   -2.07  0.0517
age        0.008  0.003    2.37  0.0280
```

```
n=25 p=5 Residual SD=0.343 R-squared=0.76
```

We see that race is significant in the north, but not in the south. By dividing the data into smaller and smaller subsets, it is possible to dilute the significance of any predictor. On the other hand, it is important not to aggregate all data without regard to whether it is reasonable. Clearly a judgment has to be made and this can be a point of contention in legal cases.

There are some special difficulties in presenting this during a court case. With scientific inquiries, there is always room for uncertainty and subtlety in presenting the results, particularly if the subject matter is not contentious. In an adversarial proceeding, it is difficult to present statistical evidence when the outcome is not clear-cut, as in this example. There are particular difficulties in explaining such evidence to non-mathematically trained people.

After all this analysis, the reader may be feeling somewhat dissatisfied. It seems we are unable to come to any truly definite conclusions, and everything we say has been hedged with "ifs" and "buts." Winston Churchill once said:

> Indeed, it has been said that democracy is the worst form of Government except all those other forms that have been tried from time to time.

We might say the same about statistics with respect to how it helps us reason in the face of uncertainty. It is not entirely satisfying, but the alternatives are worse.

Exercises

A good exercise for this chapter requires a real dataset with some practical questions of interest.

In general, a full answer requires you to perform a complete analysis of the data including an initial data anaysis, regression diagnostics, a search for possible transformations and a consideration of model selection. A report on your analysis needs to be selective in its content. You should include enough information for the steps leading to your selection of model to be clear and reproducible by the reader. But you should not include everything you tried. Dead ends can be reported in passing but do not need to be described in full detail unless they contain some message of interest. Above all, your analysis should have a clear statement of the conclusion of your analysis.

Chapter 13

Missing Data

13.1 Types of Missing Data

Here are some of the ways that missing data can arise in a regression setting:

Missing cases Sometimes we fail to observe a complete case (x_i, y_i). Indeed, when we draw a sample from a population, we do not observe the unsampled cases. When missing data arise in this manner, there is no difficulty since this is the standard situation for much of statistics. But sometimes there are cases we intended to sample but failed to observe. If the reason for this failure is unrelated to what would have been observed, then we simply have a smaller sample and can proceed as normal. But when data are not observed for reasons that have some connection to what we would have seen, then we have a biased sample. Sometimes, given enough information about the mechanism for missingness, we can make corrections and achieve valid inferences.

Incomplete values Suppose we run an experiment to study the lifetimes of light bulbs. We might run out of time waiting for all the lightbulbs to die and decide to end the experiment. These incomplete cases would provide the information that a bulb lasted at least some amount of time, but we would not know how long it would have lasted had we waited until it died. Similar examples arise in medical trials where patient final outcomes are not known. Such cases are said to be *censored*. Methods like survival analysis or reliability analysis can handle such data.

Missing values Sometimes we observe some components of a case but not others. We might observe the values of some predictors but not others. Perhaps the predictors are observed but not the response.

In this chapter, we address the missing value problem. Biased sampling due to missing cases can sometimes be mitigated by covariate adjustment, while methods for analyzing censored data may be found in other books.

What can be done? Finding the missing values is the best option, but this may not be possible because the values were never recorded or were lost in the data collection process. Next, ask why the data are missing. We can distinguish several kinds of missingness:

Missing Completely at Random (MCAR) The probability that a value is missing is the same for all cases. If we simply delete all cases with missing values from the analysis, we will cause no bias, although we may lose some information.

Missing at Random (MAR) The probability of a value is missing depends on a known mechanism. For example, in social surveys, certain groups are less likely to provide information than others. As long as we know the group membership of the individual being sampled, then this is an example of MAR. We can delete these missing cases provided we adjust for the group membership by including this as a factor in the regression model. We will see later how we may do better than simply deleting such cases.

Missing Not at Random (MNAR) The probability that a value is missing depends on some unobserved variable or, more seriously, on what value would have been observed. For example, people who have something to hide are typically less likely to provide information that might reveal something embarassing or illegal.

Dealing with the MNAR case is difficult or even impossible so we confine our attention to MAR problems.

Determining the type of missingness for data usually requires judgment alone. There are no obvious data-based diagnostic methods to check since the data we need for such checks are missing.

13.2   Representation and Detection of Missing Values

First load packages:
```
import pandas as pd
import numpy as np
import matplotlib.pyplot as plt
import statsmodels.api as sm
import statsmodels.formula.api as smf
import faraway.utils
```
Suppose some of the values in the Chicago insurance dataset of Chapter 12 were missing. I randomly declared some of the observations missing in this modified dataset:
```
import faraway.datasets.chmiss
chmiss = faraway.datasets.chmiss.load()
chmiss.index = chmiss.zip
chmiss.drop(columns=['zip'],inplace=True)
chmiss.head()
```
```
       race  fire  theft   age  involact  income
zip
60626  10.0   6.2   29.0  60.4       NaN  11.744
60640  22.2   9.5   44.0  76.5       0.1   9.323
60613  19.6  10.5   36.0   NaN       1.2   9.948
60657  17.3   7.7   37.0   NaN       0.5  10.656
60614  24.5   8.6   53.0  81.4       0.7   9.730
```
We have moved the `zip` column from a variable to the row label. The missing values are represented as `NaN`. This is appropriate for numerical missing values but Python will represent missing values of other types, for example categorical variables, with the value `None`. Be aware that different software, even other parts of Python, represents missing values differently. Furthermore, data creators may come up with their own missing value codes. For example, numerical codes such as 999 or 0 are sometimes used. Unfortunately, these can be interpreted as observed values if one is not

paying attention. It is always wise to consider if missing values may be present and how they are represented.

Sometimes, we are warned that there are missing values and we should always be alert to this possiblity in our initial data analysis. For a small dataset, simply looking at the data reveals the problem. For larger datasets, we can view the standard description

```
chmiss.describe().round(2)
```

	race	fire	theft	age	involact	income
count	43.00	45.00	43.00	42.00	44.00	45.00
mean	35.61	11.42	32.65	59.97	0.65	10.74
std	33.26	8.36	23.12	23.62	0.64	2.79
min	1.00	2.00	3.00	2.00	0.00	5.58
25%	3.75	5.60	22.00	48.30	0.00	8.56
50%	24.50	9.50	29.00	64.40	0.50	10.69
75%	57.65	15.10	38.00	78.25	0.92	12.10
max	99.70	36.20	147.00	90.10	2.20	21.48

but the only tip is the varying numbers in the count row indicating the number of complete observations. If we want the number of missing values for each variable:

```
chmiss.isna().sum(axis=0).to_frame().T
```

	race	fire	theft	age	involact	income
0	4	2	4	5	3	2

It is also helpful to see how many missing values appear in each case.

```
chmiss.isna().sum(axis=1).to_frame().T
```

zip	60626	60640	60613	60657	60614	...	60643	60628	60627	60633	60645
0	1	0	1	1	0	...	0	1	0	0	1

We see there is at most one missing value in each row. We can also plot the missing value information as seen in Figure 13.1. In this example, the missing cases are evenly scattered throughout the data. In some cases, the missing values are concentrated in some variables or cases. This can make it easier to drop variables from the analysis or delete cases without losing much information. But in this example, deleting the missing cases will lose 20 out of 47 observations.

```
plt.imshow(~chmiss.isna(), aspect='auto')
plt.xlabel("variables")
plt.xticks(np.arange(0,6), chmiss.columns)
plt.ylabel("cases")
```

13.3 Deletion

The easiest way to deal with missing values is to delete them. If the model you plan to fit has a missing value for any of the variables in a given case, you delete that case. Many statistical procedures do not handle missing values explicitly. Case deletion will allow you to fit your chosen model, provided sufficient data remain after the missing value deletion. This method is also called *complete case analysis* because only complete cases are used in the model.

Let's see what happens when we use a deletion strategy. First consider the full data fit for comparison purposes:

```
import faraway.datasets.chredlin
chredlin = faraway.datasets.chredlin.load()
lmod = smf.ols(
    'involact ~ race + fire + theft + age + np.log(income)',
```

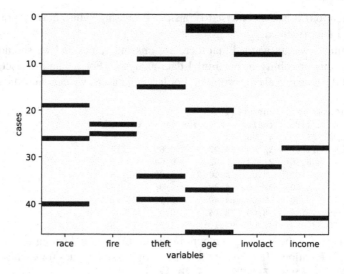

Figure 13.1   Missing values in the simulated Chicago insurance data.

```
        chredlin).fit()
lmod.sumary()
            coefs stderr tvalues pvalues
Intercept   -1.186  1.100  -1.08  0.2876
race         0.010  0.002   3.82  0.0004
fire         0.040  0.009   4.55  0.0000
theft       -0.010  0.003  -3.65  0.0007
age          0.008  0.003   3.04  0.0041
np.log(income) 0.346 0.400  0.86  0.3925

n=47 p=6 Residual SD=0.335 R-squared=0.75
```

If we really had missing values, we would not be able to see such output in practice. Now compare this to a fit to the missing version of the data:

```
lmodm = smf.ols(
    'involact ~ race + fire + theft + age + np.log(income)',
    chmiss).fit()
lmodm.sumary()
            coefs stderr tvalues pvalues
Intercept   -2.407  1.419  -1.70  0.1046
race         0.011  0.003   3.23  0.0040
fire         0.045  0.011   4.21  0.0004
theft       -0.016  0.006  -2.89  0.0087
age          0.009  0.003   2.65  0.0149
np.log(income) 0.844 0.532  1.59  0.1272

n=27 p=6 Residual SD=0.339 R-squared=0.79
```

The default behavior of the smf.ols function is to omit any case with missing values. We are left with only 27 complete cases for the regression which is the only indication that cases have been dropped (same is true for the full summary output). One can add the optional argument `missing='drop'` to the call which will cause the fit to fail. This would be wise if you have not previously checked for missing values.

We see can that the standard errors are larger for the missing version, because we have less data to fit the model and so the estimates are less precise. We can compute the ratio of the standard errors between the two models as:

```
lmodm.bse/lmod.bse
Intercept          1.289846
race               1.383112
fire               1.220096
theft              1.969479
age                1.254731
np.log(income)     1.328592
```

Given that standard errors are proportional to $\sqrt{n}$, we would expect a ratio of around $\sqrt{47/27} = 1.32$ in this example. This is roughly correct although we can see that it would not be sensible to attempt a correction to the standard errors for missing data based on the ratio. Nevertheless, it does provide us with a guide to how much precision we have lost due to missing data and whether it might be worthwhile to attempt retrieval of the missing data, collect additional data or try more sophisticated missing data methods.

Deleting missing cases is the simplest strategy for dealing with missing data. It avoids the complexity and possible biases introduced by the more sophisticated methods that we will discuss. The drawback is that we are throwing away information that might allow more precise inference. A decision on whether deletion is an acceptable missing data strategy depends on the circumstances. If relatively few cases contain missing values, if deleting missing cases still leaves a large dataset or if you wish to communicate a simple data analysis method, the deletion strategy is satisfactory.

13.4   Single Imputation

A simple solution to the problem is to fill in or *impute* the missing values. For example, we can fill in the missing values by the variable means:

```
cmeans = chmiss.mean(axis=0); cmeans.to_frame().T.round(2)
     race   fire  theft    age  involact  income
0   35.61  11.42  32.65  59.97      0.65   10.74
```

We do not fill in missing values in the response because this is the variable we are trying to model. The cases with a missing response still have some value in imputing the other missing predictor values. Now refit:

```
mchm = chmiss.copy()
mchm.race.fillna(cmeans['race'],inplace=True)
mchm.fire.fillna(cmeans['fire'],inplace=True)
mchm.theft.fillna(cmeans['theft'],inplace=True)
mchm.age.fillna(cmeans['age'],inplace=True)
mchm.income.fillna(cmeans['income'],inplace=True)
imod = smf.ols(
    'involact ~ race + fire + theft + age + np.log(income)',
    mchm).fit()
imod.sumary()
              coefs stderr tvalues pvalues
Intercept     0.530  1.088    0.49  0.6292
```

```
race                0.007  0.003    2.45  0.0189
fire                0.028  0.010    2.96  0.0053
theft              -0.003  0.003   -1.21  0.2354
age                 0.006  0.003    1.98  0.0554
np.log(income) -0.316  0.390   -0.81  0.4234
```

n=44 p=6 Residual SD=0.385 R-squared=0.68

There are some important differences between these two fits. For example, theft and age are significant in the first fit, but not in the second. Also, the regression coefficients are now all closer to zero. The situation is analogous to the errors in variables case. The bias introduced by the fill-in method can be substantial and may not be compensated by the attendant reduction in variance. For this reason, mean imputation is not recommended except where the fraction of filled values is small.

Missing values may also arise for categorical variables. We can impute with the most common level or category of the variable. Alternatively, a missing value can simply be regarded as an additional level of the variable so, for example, we might have male, female and unknown levels.

A more sophisticated alternative to mean imputation is to use regression methods to predict the missing values of the covariates. Let's try to fill in the missing race values:

```
lmodr = smf.ols(
    'race ~ fire + theft + age + np.log(income)',
    chmiss).fit()
mv = chmiss.race.isna()
lmodr.predict(chmiss)[mv]
60646   -15.945180
60651    21.418063
60616    72.607239
60617    27.977170
```

Notice that the first prediction is negative. One trick that can be applied when the response is bounded between zero and one is the logit transformation:

$$y \rightarrow \log(y/(1-y))$$

This transformation maps the interval to the whole real line. The logit function and its inverse can be defined:

```
def logit(x): return(np.log(x/(1-x)))
def ilogit(x): return(np.exp(x)/(1+np.exp(x)))
```

We now fit the model with a logit-transformed response and then back transform the predicted values, remembering to convert our percentages to proportions and vice versa at the appropriate times:

```
lmodr = smf.ols(
    'logit(race/100) ~ fire + theft + age + np.log(income)',
    chmiss).fit()
(ilogit(lmodr.predict(chmiss))*100)[mv]
60646     0.456738
60651    10.142346
60616    87.219223
60617    15.781989
```

We can see how our predicted values compare to the actual values:

```
chredlin.race.iloc[np.where(chmiss.race.isna())]
```

```
12     1.0
19    13.4
26    62.3
40    36.4
```

So our first two predictions are good, but the other two are somewhat wide of the mark. We still need to impute the missing values for the other predictors which will take us some manual effort.

Like the mean fill-in method, the regression fill-in method will also introduce a bias toward zero in the coefficients while tending to reduce the variance. The success of the regression method depends somewhat on the collinearity of the predictors — the filled-in values will be more accurate the more collinear the predictors are. However, for collinear data one might consider deleting a predictor riddled with missing values rather than imputing because other variables that are strongly correlated will bear the predictive load.

There is some difference of opinion whether to use response values in imputing missing predictor values. Given that the goal of the linear model is to predict or explain the response using the predictors, it seems unreasonable to do this.

13.5 Multiple Imputation

The single imputation methods described above cause bias, while deletion causes a loss of information from the data. Multiple imputation is a way to reduce the bias caused by single imputation. The problem with single imputation is that the imputed value, be it a mean or a regression-predicted value, tends to be less variable than the value we would have seen because the imputed value does not include the variation that would normally be seen in observed data. The idea behind multiple imputation is to reinclude that variation — we add back on a perturbation to the imputed value. *Hot deck imputation* is an old idea where a randomly sampled value from the complete values for that predictor is used as the imputation. With more computing power, we can do better than this.

We describe here the *Multiple Imputation Chained Equation* (MICE) method:

1. Use the mean single imputation method to temporarily fill the missing values as in the previous section.

2. For each variable containing missing values, we build a regression model to predict these values using the other variables. We created such a model for predicting race in the previous section. Instead of using the mean predicted value, we generate a randomly perturbed value. There are various sensible ways to do this. We could use a predictive distribution, a bootstrap-based method or by perturbing the parameter estimates using a Gaussian distribution based on the estimate standard error. The implementation below uses the latter method by default.

3. We cycle through all the variables, imputing the missing values using the perturbed samples from the model to predict that variable. Where we have categorical variables, we may need models that are not described in this book. For example, if we have a two level factor, we would need a logistic regression model.

4. At the end of one cycle through the variables, we will have replaced all of the

originally missing data. Unfortunately, the outcome will be influenced by the order in which we impute the variables. To mitigate this effect, we repeat the cycle several times. In the implementation below, this is called the *burn-in* period and is expressed as the number of skips. Ten complete cycles have been recommended by several authors.

5. Once we are past the setup period, we can generate multiple complete imputed data sets. We will want to skip through a few cycles before making a set of imputations. For each imputed set, we fit the original model and save the fitted coefficients and standard errors. The number of imputed datasets required depends on the intended use. If your focus is on parameter estimates and standard errors, about 50 should suffice. But if your dataset is not large and your model not complicated, you can easily afford to do more.

6. We now use a *combining rule* to the imputed estimates. We take the average over the imputed estimates as the point estimate. We use the estimated variances in the imputed estimates to form a single estimated standard error.

We demonstrate the method below. We have used 10 skips for burn-in and 50 imputations.

```
import statsmodels.imputation.mice as smi
imp = smi.MICEData(chmiss)
fm = 'involact ~ race + fire + theft + age + np.log(income)'
mmod = smi.MICE(fm, sm.OLS, imp)
results = mmod.fit(10, 50)
print(results.summary())
```

<pre>
                              Results: MICE
===
Method: MICE   Sample size: 47
Model: OLS   Scale   0.14
Dependent variable: involact   Num. imputations   50

                Coef.   Std.Err.   t   P>|t| [0.025 0.975] FMI

Intercept   0.2402 1.1318 0.2123 0.8319 -1.9780 2.4585 0.1066
race   0.0069 0.0028 2.4862 0.0129 0.0015 0.0123 0.0992
fire   0.0313 0.0099 3.1647 0.0016 0.0119 0.0507 0.1229
theft   -0.0046 0.0031 -1.4553 0.1456 -0.0107 0.0016 0.2976
age   0.0064 0.0030 2.1470 0.0318 0.0006 0.0123 0.0927
np.log(income) -0.1985 0.4111 -0.4829 0.6291 -1.0042 0.6072 0.1131
===
</pre>

The FMI column represents the fraction of missing information. This will be larger for predictors where the imputed variances are more variable between imputations. This indicates greater uncertainty for the effect of these predictors due to the pattern of missing values. In this case, the theft predictor is most heavily affected — we see that it is not statistically significant as it was in the complete data case.

There are many other multiple imputation methods although there is a more limited selection available with Python implementations.

13.6 Discussion

Complete case analysis is a safe choice for dealing with missing data provided there is not too high a proportion of missing data. By discarding low quality predictors with a high proportion of missing data, one can limit the proportion of cases with missing values.

Single imputation can be relatively simple to implement. It works best when there is a relatively low proportion of missing cases. But this is also a situation where complete case analysis is acceptable. The drawbacks may outweigh the advantages of this method.

Multiple imputation is the best choice for extracting the maximum of information from the data. Unfortunately, there are substantial costs both in human time in implementing the imputation models and in computing time which will be a large multiple of the effort required for a complete case dataset.

In situations where prediction rather than explanation is the goal, one must consider the possibility of missing values in the future cases where predictions will be made. A complete case approach will mean failing to predict an outcome where any of the predictors are missing. This may be unacceptable and will require some imputation. If we use multiple imputation, this will complicate the implementation particularly if the prediction will occur beyond our involvement in the project. An alternative approach is to develop a prediction for every pattern of missing values. This is called the *pattern submodel* method. If we have p predictors, this may mean up to 2^p models. When p is beyond a handful, this would limit our ability to craft the prediction models by hand.

Missing values are too big a topic to do justice in this chapter. See Little and Rubin (2002), Schafer (1997) and Raghunathan (2015) for more about missing data methods.

Exercises

1. The dataset **kanga** contains data on the skulls of historical kangaroo specimens. Ignore the sex and species variables for the purposes of this question.

   (a) Report on the distribution of missing values within the data according to case and variable.

   (b) Determine a combination of variable and case deletions that retains the most information.

   (c) Compute correlation matrix. For the variables you chose to delete in the previous part question, observe the correlation with other variables. Do you think the other variables will adequately represent the effect of the dropped variables?

   (d) Which two variables would you be most reluctant to drop?

   (e) Perform a principal components analysis using three different missing value procedures. In all cases, standardize the variables and report the first principal component. (i) Complete case analysis, (ii) Single imputation using mean value fill-in, (iii) Multiple imputation (but only do one random imputation -

you will find the `MICEData` and `next_sample` functions from the mice imputation package useful).

2. The dataset `galamiss` contains the Galapagos data frequently featured as an example in this text as `gala` but with the original missing values left in.

   (a) Fit a linear model using `gala` with the number of species as the response and
   the five geographic predictors as in earlier examples.

   (b) Fit the same model to `galamiss` using the deletion strategy for missing values.
   Compare the fit to that in (a).

   (c) Use mean value single imputation on `galamiss` and again fit the model. Compare to previous fits.

   (d) Use a regression-based imputation using the other four geographic predictors
   to fill in the missing values in `galamiss`. Fit the same model and compare to
   previous fits.

   (e) Use multiple imputation to handle missing values and fit the same model again.
   Compare to previous fits.

3. The `pima` dataset contains information on 768 adult female Pima Indians living
near Phoenix.

   (a) The analysis in Chapter 1 suggests that zero has been used as a missing value
   code for several of the variables. Replace these values with NA. Describe the
   distribution of missing values in the data.

   (b) Fit a linear model with `diastolic` as the response and the other variables as
   predictors. Summarize the fit.

   (c) Use mean value imputation to the missing cases and refit the model comparing
   to fit found in the previous question.

   (d) Use multiple imputation to handle missing values and fit the same model again.
   Compare to previous fits.

Chapter 14

Categorical Predictors

Predictors that are qualitative in nature, for example, eye color, are sometimes described as *categorical* or called *factors*. The different categories of a factor variable are called levels. For example, suppose we recognize eye colors of "blue", "green", "brown" and "hazel", then we would say eye color is a factor with four levels.

We wish to incorporate these predictors into the regression analysis. We start with the example of a factor with just two levels, then show how to introduce quantitative predictors into the model and end with an example using a factor with more than two levels.

14.1 A Two-Level Factor

The data for this example come from a study of the effects of childhood sexual abuse on adult females reported in Rodriguez et al. (1997): 45 women treated at a clinic, who reported childhood sexual abuse (`csa`), were measured for post-traumatic stress disorder (`ptsd`) and childhood physical abuse (`cpa`) both on standardized scales. Thirty-one women treated at the same clinic, who did not report childhood sexual abuse, were also measured. The full study was more complex than reported here and so readers interested in the subject matter should refer to the original article.

Load the necessary packages:
```
import pandas as pd
import numpy as np
import matplotlib.pyplot as plt
import statsmodels.api as sm
import statsmodels.formula.api as smf
import seaborn as sns
from scipy import stats
import faraway.utils
```
We take a look at the data and produce a summary subsetted by `csa`:
```
import faraway.datasets.sexab
sexab = faraway.datasets.sexab.load()
sexab.head()
```
```
      cpa      ptsd      csa
1   2.04786   9.71365   Abused
2   0.83895   6.16933   Abused
3  -0.24139  15.15926   Abused
4  -1.11461  11.31277   Abused
5   2.01468   9.95384   Abused
```
It is also helpful to have within group statistics:
```
lfuncs = ['min','median','max']
sexab.groupby('csa').agg({'cpa': lfuncs,'ptsd': lfuncs}).round(1)
```

	cpa			ptsd		
	min	median	max	min	median	max
csa						
Abused	-1.1	2.6	8.6	6.0	11.3	19.0
NotAbused	-3.1	1.3	5.0	-3.3	5.8	10.9

Now plot the data — see Figure 14.1.

```
sns.boxplot(x="csa", y="ptsd", data=sexab)
sns.pairplot(x_vars="cpa", y_vars="ptsd", data=sexab,
    hue="csa", markers=["+","o"])
```

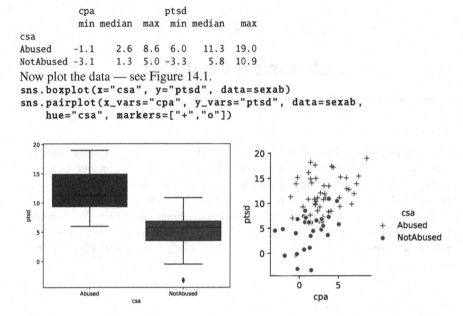

Figure 14.1   PTSD comparison of abused and non-abused subjects on the left. Relationship between PTSD and CPA is shown on the right.

We see that those in the abused group have higher levels of PTSD than those in the non-abused in the left panel of Figure 14.1. We can test this difference:

```
stats.ttest_ind(sexab.ptsd[sexab.csa == 'Abused'],
    sexab.ptsd[sexab.csa == 'NotAbused'])
Ttest_indResult(statistic=8.939, pvalue=2.172e-13)
```

and find that it is clearly significant. This test assumes that the variance is equal in the two groups. The assumption is reasonable here, but it also ensures comparability with the linear modeling to follow.

Our strategy is to incorporate qualitative predictors within the $y = X\beta + \varepsilon$ framework. We can then use the estimation, inferential and diagnostic techniques that we have already learned in a generic way. This also avoids having to learn a different set of formulae for each new type of qualitative predictor configuration.

To put qualitative predictors into the $y = X\beta + \varepsilon$ form we need to code the qualitative predictors. We can do this using *dummy variables*. For a categorical predictor (or factor) with two levels, we define dummy variables d_1 and d_2:

$$d_i = \begin{cases} 0 & \text{is not level i} \\ 1 & \text{is level i} \end{cases}$$

Let's create these dummy variables and fit them using a linear model:

```
df1 = (sexab.csa == 'Abused').astype(int)
df2 = (sexab.csa == 'NotAbused').astype(int)
X = np.column_stack((df1,df2))
lmod = sm.OLS(sexab.ptsd,sm.add_constant(X)).fit()
```

```
lmod.sumary()
      coefs stderr tvalues pvalues
const 5.546  0.270   20.53  0.0000
x1    6.395  0.403   15.87  0.0000
x2   -0.850  0.450   -1.89  0.0630
```

```
n=76 p=2 Residual SD=3.473 R-squared=0.52
Warning: Strong collinearity - design may be singular
```

We can see a warning about singularities. Although some model output is produced, its meaning is somewhat obscure under these circumstances. In particular, the NotAbused level appears not to be significant. As will be seen below, this is misleading and should be ignored. The cause of this identifiability problem can be revealed by studying the X model matrix. We show only the first, two in the middle, and last rows:

```
sm.add_constant(X)[[0,44,45,75],:]
array([[1., 1., 0.],
       [1., 1., 0.],
       [1., 0., 1.],
       [1., 0., 1.]])
```

We can see that the sum of the second and third columns equals the first column. This means that X is not of full rank, having a rank of two, not three. Hence not all the parameters can be identified. This should not be surprising since we are trying to use three parameters to model only two groups.

We have more parameters than we need, so the solution is to get rid of one of them. Once choice would be to eliminate d_1:

```
lmod = sm.OLS(sexab.ptsd,sm.add_constant(df2)).fit()
lmod.sumary()
      coefs stderr tvalues pvalues
const 11.941 0.518   23.07  0.0000
csa   -7.245 0.811   -8.94  0.0000
```

```
n=76 p=2 Residual SD=3.473 R-squared=0.52
```

Compare this to the output of the t-test. The intercept of 11.941 is the mean of the first group ("Abused") while the parameter for d2 represents the difference between the second and first group, i.e., $11.941 - 7.245 = 4.696$. The t-value for d2 of -8.94 is the test statistic for the test that the difference is zero and is identical (excepting the sign) to the test statistic from the t-test computed previously.

An alternative approach is to eliminate the intercept term:

```
lmod = sm.OLS(sexab.ptsd,X).fit()
lmod.sumary()
      coefs stderr tvalues pvalues
x1 11.941  0.518   23.07  0.0000
x2 4.696   0.624    7.53  0.0000
```

```
n=76 p=2 Residual SD=3.473 R-squared=0.52
```

The advantage of this approach is that the means of the two groups are directly supplied by the parameter estimates of the two dummy variables. However, we do not get the t-test for the difference. The tests in the output correspond to hypotheses claiming the mean response in the group is zero. These are not interesting because these hypotheses are unbelievable.

Furthermore, the solution of dropping the intercept only works when there is a single factor and does not generalize to the multiple factor case. For these reasons, we prefer the approach of dropping one of the dummy variables to dropping the intercept.

It is not necessary to explicitly form the dummy variables as statsmodels via patsy can produce these directly by just including the factor in the model formula:

```
lmod = smf.ols('ptsd ~ csa', sexab).fit()
lmod.sumary()
```

```
                 coefs stderr tvalues pvalues
Intercept        11.941 0.518  23.07   0.0000
csa[T.NotAbused] -7.245 0.811  -8.94   0.0000
```

n=76 p=2 Residual SD=3.473 R-squared=0.52

We can check that csa is a factor variable:

```
sexab.csa.dtype, sexab.ptsd.dtype
(dtype('O'), dtype('float64'))
```

This happens automatically when a variable takes non-numeric values. The dummy variables are created but one is dropped to ensure identifiability. This is known as the *reference level*. In this example, the reference level is "Abused". At first glance, one might be perplexed as "Abused" does not appear in the model summary output. However, the mean response for the reference level is encoded in the intercept of 11.941. The parameter estimate for "NotAbused" of -7.245 is not the mean response for this level but the difference from the reference level. Hence the mean response for the "NotAbused" level is $11.941 - 7.245 = 4.696$. Earlier in our analysis, we dropped the dummy variable d_1 to get the same output.

We can also create dummy variables in pandas:

```
sac = pd.concat([sexab,pd.get_dummies(sexab.csa)],axis=1)
sac.iloc[[0,44,45,75],:]
```

```
        cpa      ptsd       csa  Abused  NotAbused
1   2.04786   9.71365    Abused       1          0
45  5.11921  11.12798    Abused       1          0
46  1.49181   6.14200 NotAbused       0          1
76  0.81138   7.12918 NotAbused       0          1
```

and fit the model:

```
lmod = smf.ols('ptsd ~ Abused', sac).fit()
lmod.sumary()
```

```
          coefs stderr tvalues pvalues
Intercept 4.696  0.624   7.53   0.0000
Abused    7.245  0.811   8.94   0.0000
```

n=76 p=2 Residual SD=3.473 R-squared=0.52

The choice of reference level is arbitrary. In some examples, there is a natural choice for the reference level as a control or no-treatment level. In the current example, "NotAbused" is the natural choice for the reference level. The default choice of reference level is the first level in alphabetical order, which would be "Abused" in this example. Because this choice is inconvenient, we change the reference level using the relevel command:

```
lmod = smf.ols(
    'ptsd ~ C(csa,Treatment(reference="NotAbused"))',
    sexab).fit()
lmod.sumary()
```

```
                                                 coefs stderr tvalues pvalues
Intercept                                        4.696 0.624    7.53  0.0000
C(csa, Treatment(reference="NotAbused"))[T.Abused] 7.245 0.811  8.94  0.0000
```

n=76 p=2 Residual SD=3.473 R-squared=0.52

A comparison of the outputs reveals that the fitted values and residuals are the same for either choice — the residual standard error and R^2 will be the same. But the parameterization is different.

Although we have only managed to construct a t-test using linear modeling, a good understanding of how factor variables are handled is essential for the more sophisticated models to follow.

14.2 Factors and Quantitative Predictors

We can see from our analysis that women who have suffered childhood sexual abuse tend to have higher levels of PTSD than those who have not. However, we also have information about varying levels of childhood physical abuse (cpa) which may also have an effect on PTSD. We need models that can express how a factor variable like csa and a quantitative variable like cpa might be related to a response.

Suppose we have a response y, a quantitative predictor x and a two-level factor variable represented by a dummy variable d:

$$d = \begin{cases} 0 & \text{reference level} \\ 1 & \text{treatment level} \end{cases}$$

Several possible linear models may be considered here:

1. The same regression line for both levels: $y = \beta_0 + \beta_1 x + \varepsilon$ or is written as y ~ x. This model allows no effect for the factor.

2. A factor predictor but no quantitative predictor: $y = \beta_0 + \beta_2 d + \varepsilon$. This is written as y ~ d.

3. Separate regression lines for each group with the same slope: $y = \beta_0 + \beta_1 x + \beta_2 d + \varepsilon$ or is written as y ~ x + d. In this case β_2 represents the vertical distance between the regression lines (i.e., the effect of the treatment).

4. Separate regression lines for each group with the different slopes: $y = \beta_0 + \beta_1 x + \beta_2 d + \beta_3 x.d + \varepsilon$ or is written as y ~ x + d + d:x or y ~ x*d. To form the slope interaction term d:x in the X-matrix, multiply x by d elementwise. Any interpretation of the effect of the factor will now also depend on the quantitative predictor.

Estimation and testing work just as they did before. Interpretation is easier if we can eliminate the interaction term.

We start with the separate regression lines model — ptsd ~ cpa*csa:
```
lmod4 = smf.ols('ptsd ~ csa*cpa', sexab).fit()
lmod4.sumary()
                coefs stderr tvalues pvalues
Intercept      10.557  0.806  13.09  0.0000
csa[T.NotAbused] -6.861 1.075 -6.38  0.0000
cpa             0.450  0.208   2.16  0.0342
```

```
csa[T.NotAbused]:cpa  0.314  0.368     0.85  0.3970
```

```
n=76 p=4 Residual SD=3.279 R-squared=0.58
```
We can discover the coding by examining the X-matrix:
```
import patsy
patsy.dmatrix('~ csa*cpa', sexab)[[0,44,45,75],]
array([[1.    , 0.    , 2.0479, 0.    ],
       [1.    , 0.    , 5.1192, 0.    ],
       [1.    , 1.    , 1.4918, 1.4918],
       [1.    , 1.    , 0.8114, 0.8114]])
```
The interaction term cpa:csaAbused is represented in the fourth column of the matrix as the product of the second and third columns. It represents the change in slope for the notAbused group. We show the fitted regression lines in the left panel of Figure 14.2:
```
abused = (sexab.csa == "Abused")
plt.scatter(sexab.cpa[abused], sexab.ptsd[abused], marker='x')
xl,xu = [-3, 9]
a, b = (lmod4.params[0], lmod4.params[2])
plt.plot([xl,xu], [a+xl*b,a+xu*b])
plt.scatter(sexab.cpa[~abused], sexab.ptsd[~abused], marker='o')
a, b = (lmod4.params[0]+lmod4.params[1],
        lmod4.params[2]+lmod4.params[3])
plt.plot([xl,xu], [a+xl*b,a+xu*b])
```
The model can be simplified because the interaction term is not significant. We claim the slopes are the same. We reduce to this model:
```
lmod3 = smf.ols('ptsd ~ csa+cpa', sexab).fit()
lmod3.sumary()
                 coefs stderr tvalues pvalues
Intercept       10.248  0.719   14.26  0.0000
csa[T.NotAbused] -6.273  0.822   -7.63  0.0000
cpa              0.551  0.172    3.21  0.0020
```

```
n=76 p=3 Residual SD=3.273 R-squared=0.58
```
No further simplification is possible because the remaining predictors are statistically significant. Put the two parallel regression lines on the plot, as seen in the middle panel of Figure 14.2.
```
plt.scatter(sexab.cpa[abused], sexab.ptsd[abused], marker='x')
xl,xu = [-3, 9]
a, b = (lmod3.params[0], lmod3.params[2])
plt.plot([xl,xu], [a+xl*b,a+xu*b])
plt.scatter(sexab.cpa[~abused], sexab.ptsd[~abused], marker='o')
a, b = (lmod3.params[0]+lmod4.params[1], lmod3.params[2])
plt.plot([xl,xu], [a+xl*b,a+xu*b])
```
The slope of both lines is 0.551, but the "Abused" line is 6.273 higher than the "NonAbused." From the t-test earlier, the unadjusted estimated effect of childhood sexual abuse is 7.245. So after adjusting for the effect of childhood physical abuse, our estimate of the effect of childhood sexual abuse on PTSD is mildly reduced.

We can also compare confidence intervals for the effect of csa:
```
lmod3.conf_int().round(2)
                   0      1
Intercept        8.82  11.68
csa[T.NotAbused] -7.91  -4.63
cpa              0.21   0.89
```

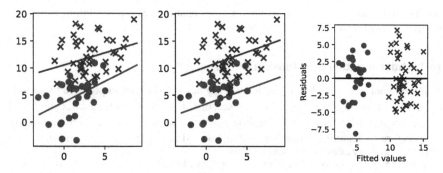

Figure 14.2 In all three plots, 'x' marks the abused group, while dot represents the not abused group. The separate lines model fit is shown in the left panel, the parallel lines model in the middle panel and the residuals versus fitted plot in the right.

compared to the $(5.6302, 8.8603)$ found for the unadjusted difference. In this particular case, the confidence intervals are about the same width. In other cases, particularly with designed experiments, adjusting for a covariate can increase the precision of the estimate of an effect.

The usual diagnostics should be checked. It is worth checking whether there is some difference related to the categorical variable as we do here:

```
plt.scatter(lmod3.fittedvalues[abused], lmod3.resid[abused],
    marker='x')
plt.scatter(lmod3.fittedvalues[~abused], lmod3.resid[~abused],
    marker='o')
plt.xlabel("Fitted values")
plt.ylabel("Residuals")
plt.axhline(0)
```

We see in the right panel of Figure 14.2 that there are no clear problems. Furthermore, because the two groups happen to separate, we can also see that the variation in the two groups is about the same. If this were not so, we would need to make some adjustments to the analysis, possibly using weights.

We have seen that the effect of csa can be adjusted for cpa. The reverse is also true. Consider a model with just cpa.

```
lmod1 = smf.ols('ptsd ~ cpa', sexab).fit()
lmod1.sumary()
          coefs stderr tvalues pvalues
Intercept 6.552  0.707   9.26  0.0000
cpa       1.033  0.212   4.87  0.0000

n=76 p=2 Residual SD=4.359 R-squared=0.24
```

After adjusting for the effect of csa, we see size of the effect of cpa is reduced from 1.033 to 0.551.

Finally, we should point out that childhood physical abuse might not be the only factor that is relevant to assessing the effects of childhood sexual abuse. It is quite possible that the two groups differ according to other variables such as socioeconomic status and age. Furthermore, we must regard all causal conclusions from

observational studies with the usual sense of caution. See the original article by
Rodriguez et al. (1997) for more about this.

14.3   Interpretation with Interaction Terms

A homeowner in England recorded his weekly natural gas consumption, in thousands
of cubic feet, during two winter heating seasons. For the second season, cavity wall
insulation had been installed. The homeowner also recorded the average weekly
temperature in degrees Celsius because this would also affect gas consumption.

```
import faraway.datasets.whiteside
whiteside = faraway.datasets.whiteside.load()
```

We plot the data in Figure 14.3:

```
sns.lmplot(x="Temp", y="Gas", col="Insul", data=whiteside)
```

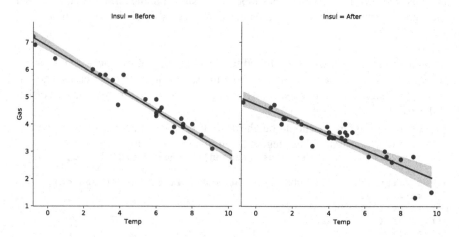

Figure 14.3   Weekly gas consumption as a function of weekly temperature before
and after the installation of cavity wall insulation.

We can see that less gas is used after the insulation is installed, but the difference
varies by temperature. The relationships appear linear.

Before we fit a model, it is helpful to specify the reference level for Insul. By
default, this would be the After level because this is first alphabetically. But it is
more natural to have Before as the reference level since we would want to compare
to this level. We can achieve this effect most conveniently with:

```
whiteside['Insul'] = pd.Categorical(whiteside['Insul'],
                        categories=['Before','After'])
whiteside['Insul'].dtype
CategoricalDtype(categories=['Before', 'After'], ordered=False)
```

This sets up a Categorical type variable where we have specified the ordering of
the levels with Before coming first. This will be respected when fitting the subsequent models.

We start with a model with two separate:

```
lmod = smf.ols('Gas ~ Temp*Insul', whiteside).fit()
lmod.summary()
```

	coefs	stderr	tvalues	pvalues
Intercept	6.854	0.136	50.41	0.0000
Insul[T.After]	-2.130	0.180	-11.83	0.0000
Temp	-0.393	0.022	-17.49	0.0000
Temp:Insul[T.After]	0.115	0.032	3.59	0.0007

n=56 p=4 Residual SD=0.323 R-squared=0.93

We would predict that the gas consumption would fall by 0.393 for each 1°C increase in temperature before insulation. After insulation, the fall in consumption per degree is only $0.393 - 0.115 = 0.278$. The interpretation for the other two parameter estimates represent predicted consumption when the temperature is zero. Gas consumption before at this temperature is 6.854 and 6.854-2.130 after. This is on the lower edge of the observed range of temperatures and would not represent a typical difference. For other datasets, a continuous predictor value of zero might be far outside the range and so these parameters would have little practical meaning.

The solution is to center the temperature predictor by its mean value and recompute the linear model:

```
whiteside.Temp.mean()
4.875
whiteside['cTemp'] = whiteside.Temp - whiteside.Temp.mean()
lmod = smf.ols('Gas ~ cTemp*Insul', whiteside).fit()
lmod.sumary()
```

	coefs	stderr	tvalues	pvalues
Intercept	4.937	0.064	76.85	0.0000
Insul[T.After]	-1.568	0.088	-17.87	0.0000
cTemp	-0.393	0.022	-17.49	0.0000
cTemp:Insul[T.After]	0.115	0.032	3.59	0.0007

n=56 p=4 Residual SD=0.323 R-squared=0.93

Now we can say that the average consumption before insulation at the average temperature was 4.94 and $4.94 - 1.57 = 3.37$ afterwards. The other two coefficients are unchanged and their interpretation remains the same. Thus we can see that centering allows a more typical interpretation of the parameter estimates in the presence of interaction.

We would not expect the installation to insulation to have constant absolute effect on the use of fuel. As we might expect, the gains are higher at lower temperatures. A significant interaction effect is what we expect. Instead, we might expect the insulation to have a relative effect on the use of fuel. We can achieve this with a log transformation on the response:

```
lmod = smf.ols('np.log(Gas) ~ Temp*Insul', whiteside).fit()
lmod.sumary()
```

	coefs	stderr	tvalues	pvalues
Intercept	1.968	0.054	36.46	0.0000
Insul[T.After]	-0.340	0.072	-4.75	0.0000
Temp	-0.082	0.009	-9.20	0.0000
Temp:Insul[T.After]	-0.011	0.013	-0.85	0.3983

n=56 p=4 Residual SD=0.128 R-squared=0.84

Now the interaction term is not significant and can be removed:

```
lmod = smf.ols('np.log(Gas) ~ Temp+Insul', whiteside).fit()
lmod.sumary()
```

```
                coefs stderr tvalues pvalues
Intercept       1.997  0.042   47.24  0.0000
Insul[T.After] -0.393  0.035  -11.31  0.0000
Temp           -0.087  0.006  -13.75  0.0000
```

n=56 p=3 Residual SD=0.128 R-squared=0.84

The nonrandom part of this model states that:

$$\log \text{Gas} = \beta_0 + \beta_{Insul} + \beta_{temp}\text{Temp}$$

which we can write as:

$$\text{Gas} = e^{\beta_0} e^{\beta_{Insul}} e^{\beta_{temp}\text{Temp}}$$

Hence the effect of the insulation is to multiply the response by $e^{-0.393} = 0.675$. So a readily understood conclusion of this analysis is that the effect of insulation is to reduce gas consumption by 32.5%.

14.4 Factors with More Than Two Levels

Suppose we have a factor with f levels, then we create $f - 1$ dummy variables $d_2, \ldots, d_f$ where:

$$d_i = \begin{cases} 0 & \text{is not level i} \\ 1 & \text{is level i} \end{cases}$$

Level one is the reference level and would be dropped to ensure identifiability.

We demonstrate the use of multilevel factors with a study on the sexual activity and the life span of male fruitflies by Partridge and Farquhar (1981): 125 fruitflies were divided randomly into five groups of 25 each. The response was the longevity of the fruitfly in days. One group was kept solitary, while another was kept individually with a virgin female each day. Another group was given eight virgin females per day. As an additional control, the fourth and fifth groups were kept with one or eight pregnant females per day. Pregnant fruitflies will not mate. The thorax length of each male was measured as this was known to affect longevity. The five groups are labeled isolated, low, high, one and many, respectively. The purpose of the analysis is to determine the difference between the five groups if any. We start with a plot of the data, as seen in Figure 14.4.

```
import faraway.datasets.fruitfly
ff = faraway.datasets.fruitfly.load()
sns.pairplot(x_vars="thorax", y_vars="longevity", data=ff,
    hue="activity",height=5,markers=["o",".","+","x","v"])
```

With multiple levels, it can be hard to distinguish the groups. Sometimes it is better to plot each level separately. This can be achieved nicely with the help of the seaborn package:

```
sns.lmplot(x="thorax", y="longevity", data=ff,
    col="activity",height=5,col_wrap=3)
```

The plot, shown in Figure 14.5, makes it clearer that longevity for the high sexual activity group is lower.

We fit and summarize the most general linear model:

```
lmod = smf.ols('longevity ~ thorax*activity', ff).fit()
lmod.sumary()
```

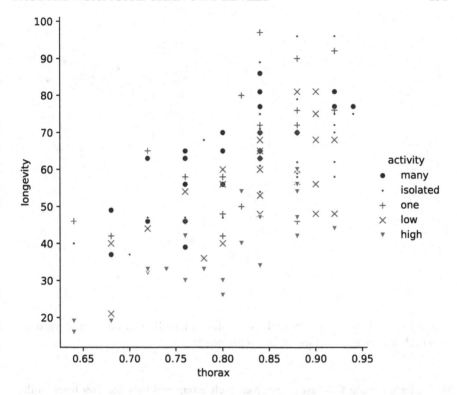

Figure 14.4   Plot of longevity in days and thorax length in millimeters of fruitflies divided into five treatment groups. Longevity for the high sexual activity group appears to be lower.

	coefs	stderr	tvalues	pvalues
Intercept	-50.242	21.801	-2.30	0.0230
activity[T.low]	-7.750	33.969	-0.23	0.8199
activity[T.high]	-11.038	31.287	-0.35	0.7249
activity[T.one]	6.517	33.871	0.19	0.8478
activity[T.many]	-1.139	32.530	-0.04	0.9721
thorax	136.127	25.952	5.25	0.0000
thorax:activity[T.low]	0.874	40.425	0.02	0.9828
thorax:activity[T.high]	-11.127	38.120	-0.29	0.7709
thorax:activity[T.one]	-4.677	40.652	-0.12	0.9086
thorax:activity[T.many]	6.548	39.360	0.17	0.8682

n=124 p=10 Residual SD=10.713 R-squared=0.65

Since "isolated" is the reference level, the fitted regression line within this group is longevity= $-50.2 + 136.1*$thorax. For "many," it is longevity= $(-50.2 - 1.1) + (136.1 + 6.5)*$thorax. Similar calculations can be made for the other groups. We can take a closer look at the design matrix to see how the variables are coded.

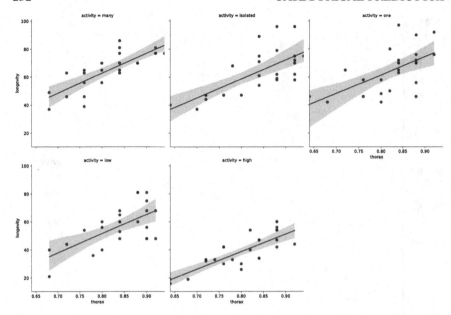

Figure 14.5   Plot of longevity in days and thorax length in millimeters of fruitflies
with each treatment group shown in a separate panel.

Here I have chosen five cases, one from each group and only the low level so the
matrix is not too big:

```
mm = patsy.dmatrix('~ thorax*activity', ff)
ii = (1, 25, 49, 75, 99)
p = pd.DataFrame(mm[ii,:],index=ii,columns=lmod.params.index)
p.iloc[:,[0,1,5,6]]
```

	Intercept	activity[T.low]	thorax	thorax:activity[T.low]
1	1.0	0.0	0.68	0.00
25	1.0	0.0	0.70	0.00
49	1.0	0.0	0.64	0.00
75	1.0	1.0	0.68	0.68
99	1.0	0.0	0.64	0.00

For the low case (fourth line), we can see the coding differs from the other levels.

There is perhaps some heteroscedasticity, but we will let this be until later for
ease of presentation. Now we see whether the model can be simplified. The model
summary output is not suitable for this purpose because there are four t-tests corresponding to the interaction term while we want just a single test for this term. We
can obtain this using:

```
sm.stats.anova_lm(lmod)
```

	df	sum_sq	mean_sq	F	PR(>F)
activity	4.0	12269.47	3067.37	26.73	0.00
thorax	1.0	12368.42	12368.42	107.77	0.00
thorax:activity	4.0	24.31	6.08	0.05	0.99
Residual	114.0	13082.98	114.76	NaN	NaN

This is a sequential analysis of variance (ANOVA) table. Starting from a null model, terms are added and sequentially tested. The models representing the null and alternatives are listed in Table 14.1. We wish to successively simplify the full model and

Null	Alternative
y~1	y~thorax
y~thorax	y~thorax+activity
y~thorax+activity	y~thorax+activity+thorax:activity

Table 14.1   Models compared in the sequential ANOVA.

then interpret the result. The interaction term thorax:activity is not significant, indicating that we can fit the same slope within each group. No further simplification is possible.

We notice that the F-statistic for the test of the interaction term is very small and its p-value close to one. For these data, the fitted regression lines to the five groups happen to be very close to parallel. This can, of course, just happen by chance. In some other cases, unusually large p-values have been used as evidence that data have been tampered with or "cleaned" to improve the fit. Most famously, Ronald Fisher suspected Gregor Mendel of fixing the data in some genetics experiments because the data seemed too good to be true. See Fisher (1936).

We now refit without the interaction term:
```
lmod = smf.ols('longevity ~ thorax+activity', ff).fit()
lmod.sumary()
```
	coefs	stderr	tvalues	pvalues
Intercept	-48.749	10.850	-4.49	0.0000
activity[T.low]	-7.015	2.981	-2.35	0.0203
activity[T.high]	-20.004	3.016	-6.63	0.0000
activity[T.one]	2.637	2.984	0.88	0.3786
activity[T.many]	4.139	3.027	1.37	0.1741
thorax	134.341	12.731	10.55	0.0000

n=124 p=6 Residual SD=10.539 R-squared=0.65

Do we need both thorax and activity? We could use the output above which suggests both terms are significant.

"Isolated" is the reference level. We see that the intercepts of "one" and "many" are not significantly different from this reference level. We also see that the low sexual activity group, "low," survives about seven days less. The p-value is 0.02 and is enough for statistical significance if only one comparison is made. However, we are making more than one comparison, and so, as with outliers, a Bonferroni-type adjustment might be considered. This would erase the statistical significance of the difference. However, the high sexual activity group, "high," has a life span 20 days less than the reference group and this is strongly significant.

Returning to the diagnostics:
```
g = sns.residplot(lmod.fittedvalues, lmod.resid)
g.set(xlabel="Fitted values",ylabel="Residuals")
```
is seen in the first panel of Figure 14.6. We have some nonconstant variance although it does not appear to be related to the five groups. A log transformation can remove the heteroscedasticity:

```
lmod = smf.ols('np.log(longevity) ~ thorax+activity', ff).fit()
g = sns.residplot(lmod.fittedvalues, lmod.resid)
g.set(xlabel="Fitted values",ylabel="Residuals")
```
as seen in the second panel of Figure 14.6. One disadvantage of transformation is

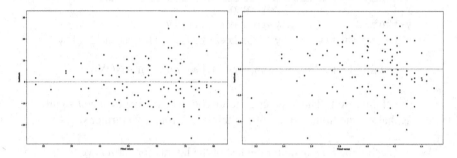

Figure 14.6   Diagnostic plots for the fruitfly data before and after log transformation
of the response.

that it can make interpretation of the model more difficult. Let's examine the model
fit:

```
                  coefs stderr tvalues pvalues
Intercept       -48.749 10.850   -4.49  0.0000
activity[T.low]  -7.015  2.981   -2.35  0.0203
activity[T.high] -20.004 3.016   -6.63  0.0000
activity[T.one]    2.637 2.984    0.88  0.3786
activity[T.many]   4.139 3.027    1.37  0.1741
thorax           134.341 12.731  10.55  0.0000
```

```
n=124 p=6 Residual SD=10.539 R-squared=0.65
```
Notice that the R^2 is higher for this model, but the p-values are similar. Because of
the log transformation, we can interpret the coefficients as having a multiplicative
effect:
```
np.round(np.exp(lmod.params[1:5]),2)
activity[T.low]    0.88
activity[T.high]   0.66
activity[T.one]    1.05
activity[T.many]   1.09
```
Compared to the reference level, we see that the high sexual activity group has 0.66
times the life span (i.e., 34% less).

Why did we include thorax in the model? Its effect on longevity was known,
but because of the random assignment of the flies to the groups, this variable will
not bias the estimates of the effects of the activities. We can verify that thorax is
unrelated to the activities:
```
lmod = smf.ols('thorax ~ activity', ff).fit()
sm.stats.anova_lm(lmod)
            df sum_sq mean_sq     F  PR(>F)
activity   4.0  0.026   0.006 1.109   0.355
Residual 119.0  0.685   0.006   NaN     NaN
```
However, look what happens if we omit thorax from the model for longevity:
```
lmod = smf.ols('np.log(longevity) ~ activity', ff).fit()
```

```
lmod.sumary()
              coefs stderr tvalues pvalues
Intercept     4.119  0.056   72.99  0.0000
activity[T.low]  -0.120  0.080   -1.50  0.1370
activity[T.high] -0.517  0.080   -6.48  0.0000
activity[T.one]   0.023  0.080    0.29  0.7695
activity[T.many]  0.024  0.081    0.30  0.7670
```

```
n=124 p=5 Residual SD=0.282 R-squared=0.36
```
The magnitudes of the effects do not change that much, but the standard errors are substantially larger. The value of including thorax in this model is to increase the precision of the estimates.

14.5   Alternative Codings of Qualitative Predictors

Sometimes an alternative coding of factor variables can be useful. Consider the previous model where we used the default treatment coding. Another choice is the *sum* coding:
```
smod = smf.ols('np.log(longevity) ~ C(activity,Sum)', ff).fit()
smod.sumary()
                          coefs stderr tvalues pvalues
Intercept                 4.001  0.025  157.88  0.0000
C(activity, Sum)[S.isolated]  0.118  0.051    2.33  0.0214
C(activity, Sum)[S.low]      -0.002  0.051   -0.03  0.9741
C(activity, Sum)[S.high]     -0.399  0.051   -7.90  0.0000
C(activity, Sum)[S.one]       0.141  0.051    2.80  0.0060
```

```
n=124 p=5 Residual SD=0.282 R-squared=0.36
```
The fit of this model is the same — statistics such as the residual standard error and R^2 are identical, but the parameter estimates are different. In the treatment coding, we set a reference level, in this case "isolated". Using the previous model output, the estimated effect for this level is given by the intercept term, 4.119, while the effect for "low" is given by 4.119-0.120=3.999. For the sum coding, the intercept represents the mean of the mean effects (so for a balanced design, this would be the overall mean). In this case, this value is 4.001. For the "low" level, we compute the difference from this, 4.001-0.002=3.999. Hence the estimated effects are the same for both codings but the parameterization is different.

Under the sum coding, the factor effects sum to zero. To avoid identifiability problems, the dummy variable corresponding to the last level is omitted. For this reason, we do not see a parameter estimate for many in the model output. But we can compute the effect for this level by using the sum to zero constraint. We have:
```
np.sum(smod.params[1:])
-0.142
```
So the estimated effect for "many" is 0.142 above the overall mean.

In general, let B be an $n \times k$ dummy variable matrix where $B_{ij} = 1$ if case i falls in class j and is zero otherwise. We might use B to form part of the model matrix. However, the row sums of B are all one. Since an intercept term would also be represented by a column of ones, all the parameters would not be identifiable.

The coding is then determined by a *contrast matrix* C which has dimension $k \times$

$(k-1)$. The contribution to the model matrix is then given by BC. Other columns of the model matrix might include a column of ones for the intercept and perhaps other predictors.

Treatment coding

Consider a five-level factor that will be coded using four dummy variables. This contrast matrix describes the coding, where the columns represent the dummy variables and the rows represent the levels:

```
from patsy.contrasts import Treatment
levels = [1,2,3,4]
contrast = Treatment(reference=0).code_without_intercept(levels)
print(contrast.matrix)
[[0. 0. 0. 0.]
 [1. 0. 0. 0.]
 [0. 1. 0. 0.]
 [0. 0. 1. 0.]
 [0. 0. 0. 1.]]
```

This treats the first level as the standard level to which all other levels are compared so a control group, if one exists, would be appropriate for this level. The parameter for the dummy variable then represents the difference between the given level and the first level. The levels are assigned to a factor in alphabetical order by default.

Sum coding

This coding uses a contrast matrix of the form:

```
from patsy.contrasts import Sum
contrast = Sum().code_without_intercept(levels)
print(contrast.matrix)
[[ 1.  0.  0.  0.]
 [ 0.  1.  0.  0.]
 [ 0.  0.  1.  0.]
 [ 0.  0.  0.  1.]
 [-1. -1. -1. -1.]]
```

We can see five representative rows of the design matrix for the last fitted model:

```
mm = patsy.dmatrix('~ C(activity,Sum)', ff)
ii = [1, 25, 49, 75, 99]
pd.DataFrame(mm[ii,:], index=ff.activity.iloc[ii],
    columns=['intercept','isolated','low','high','one'])
```

activity	intercept	isolated	low	high	one
many	1.0	-1.0	-1.0	-1.0	-1.0
isolated	1.0	1.0	0.0	0.0	0.0
one	1.0	0.0	0.0	0.0	1.0
low	1.0	0.0	1.0	0.0	0.0
high	1.0	0.0	0.0	1.0	0.0

There are other choices of coding — anything that spans the $k-1$ dimensional space will work. The Helmert and difference codings are readily available in patsy but anything can be coded by specifying the contrast matrix. The choice of coding does not affect the R^2, $\hat{\sigma}^2$ and overall F-statistic. It does affect the $\hat{\beta}$ and you do need to know what the coding is before making conclusions about $\hat{\beta}$.

Exercises

1. This question uses the `teengamb` data with `gamble` as the response and the other variables as possible predictors.

   (a) Considering just `income` and `sex` as potential predictors, make plots of the data in two distinct ways. Judging from just the plots, do you think there is an interaction effect? Why?

   (b) Fit a model allowing for an interaction effect. Is the interaction significant?

   (c) What is the interpretation of the `sex` parameter in this model? It is not statistically significant. What does this indicate?

   (d) The data already has `sex` as a dummy variable. Create a factor where the levels are labelled male and female and where female is the reference level. Now fit the model again. What does this model say about the relationship between income and gambling for females?

   (e) Fit a model which checks for an interaction between sex and each of three quantitative predictors: income, status and verbal.

   (f) Should we prefer this model to the previous simpler model?

2. The `infmort` data shows the relationship between infant mortality and income in different regions of the world.

   (a) Make a plot showing the relationship between income and infant mortality. Show the region on different facets and the oil variables with a different plotting color. Use a log scale for the income.

   (b) Which country is a clear outlier?

   (c) Fit a model which includes interactions between income and oil, and, income and region. Include all the data.

   (d) Which interactions are statistically significant?

   (e) Repeat the model fitting and testing but with the outlier excluded. What difference does this make?

   (f) Interpret the chosen model. In particular, does infant mortality increase or decrease with income? Also is infant mortality in Africa higher because of lower incomes?

3. Obtain the ToothGrowth data with
   ```
   tgdf = sm.datasets.get_rdataset("ToothGrowth")
   tg = tgdf.data
   ```
   and documentation with
   ```
   print(tgdf.__doc__)
   ```

   (a) Plot the data with `len` as the response, `dose` as the predictor and with `supp` distinguished in the plotting character.

   (b) Fit a model with `len` as the response and with `dose` and `supp` as predictors along with their interaction. Can this model be simplified?

   (c) Make a residual-fitted plot. Does this show any problems?

   (d) Fit a model where there is a quadratic term in `dose` as well as interaction(s) with `supp`. Is there are difference in curvature between the two groups?

(e) Use an F-test to compare this to the previous model. What does the outcome of the test mean?

(f) Compute the group means by dose and supp together. What does this suggest?

4. In this question, we investigate whether there is any difference between the north and south sides of Chicago in the chredlin data.

   (a) Fit a model with involact as the response and the predictors, race, fire, theft, income(logged) interacting with side. Explain what the interaction terms represent in this situation.

   (b) Test whether any of the interaction effects are significant.

   (c) Is there evidence of any effect due to side?

   (d) Fit a model with the five predictors for just the south side. Fit the same model for the north side. How can we test whether the coefficients in these two models are the same?

5. Use the uswages data with wages as the response.

   (a) Fit a model with wage as the response and the four regional indicators as predictors. Explain why a warning message is seen in the model output.

   (b) Fit the same model but without an intercept term. Why is there no error message now? Verify that the coefficient for the northeast is equal to the mean wage in the northeast. Demonstrate how the same number can be calculated from the model in (a).

   (c) Make a plot where wage (on a log scale) is the response and education is the predictor. Distinguish the race using the plotting character. How effective is this plot?

   (d) Make a plot where two boxplots for each level of race appear for each number of years of education. How does this compare to the previous plot?

   (e) Fit a model where logged wages is the response. Use an interaction between race and education. Which terms are not statistically significant?

   (f) Now fit a model where ony race is used as a predictor. Use an F-test to compare this model to the previous model.

   (g) Fit the best model - it may be one of the two above or a third model.

6. The dataset clot contains the clotting times of blood varying as percentage concentration of prothrombin-free plasma. There are two lots of thromboplastin.

   (a) Plot the data with time as the response using a different plotting symbol according to the lot.

   (b) Find the transformation of the two continuous variables to form a linear relationship.

   (c) Does the time to clot vary according to concentration differently in the two lots?

   (d) Check the assumptions of your model using regression diagnostics.

   (e) At what percentage concentration is the predicted time the same for the two lots?

7. The wealth in billions of dollars for 232 billionaires is given in fortune.

   (a) Plot the wealth as a function of age using a different plotting symbol for the different regions of the world.

   (b) Plot the wealth as a function of age with a separate panel for each region.

   (c) Determine a transformation on the response to facilitate linear modeling.

   (d) What is the relationship of age and region to wealth?

   (e) Check the assumptions of your model using appropriate diagnostics.

8. Ankylosing spondylitis is a chronic form of arthritis. A study was conducted to determine whether daily stretching of the hip tissues would improve mobility. The data are found in hips. The flexion angle of the hip before the study is a predictor, and the flexion angle after the study is the response.

   (a) Plot the data using different plotting symbols for the treatment and the control status.

   (b) Fit a model to determine whether there is a treatment effect.

   (c) Compute the difference between the flexion before and after and test whether this difference varies between treatment and control. Contrast this approach to your previous model.

   (d) Check for outliers. Explain why we might remove the three cases with fbef less than 90. Refit an appropriate model and check for a treatment effect.

   (e) What is the estimated size of the treatment effect? Give a 95% confidence interval.

   (f) Both legs of each subject have been included in the study as separate observations. Explain what difficulties this causes with the model assumptions.

   (g) Compute the average angles for each subject and repeat the modeling with this reduced dataset. Point out any differences in the conclusions.

Chapter 15

One-Factor Models

In the models considered previously, we have always had at least one quantitative predictor. For the remainder of the book, we focus on qualitative predictors. In this chapter, we consider only one qualitative predictor. Although this may seem like a simple situation, we shall see it holds some interest as well as forming the basis for the chapters to follow.

Linear models with only categorical predictors (or factors) have traditionally been called analysis of variance (ANOVA) problems. The idea is to partition the overall variance in the response due to each of the factors and the error. This traditional approach is exemplified by Scheffé (1959).

Some insight can be gained by considering the problem from this perspective, but historically this required an increasingly complex set of specialized formulae for each type of model depending on the configuration of factors. Unbalanced designs due to missing or observational data caused a particular difficulty. We avoid these problems by continuing with our current approach by putting the model into the $y = X\beta + \varepsilon$ format and then using the inferential methods we have already developed.

The terminology used in ANOVA-type problems is sometimes different. Predictors are now all qualitative and are now typically called *factors*, which have some number of *levels*. The regression parameters are now often called *effects*. We shall consider only models where the parameters are considered fixed, but unknown — called *fixed-effects* models. *Random-effects* models are used where parameters are taken to be random variables and are not covered in this text.

15.1 The Model

Suppose we have a factor α occurring at $i = 1, \ldots, I$ levels, with $j = 1, \ldots, J_i$ observations per level. We use the model:

$$y_{ij} = \mu + \alpha_i + \varepsilon_{ij}$$

The parameters are not identifiable. For example, we could add some constant to μ and subtract the same constant from each α_i and the fit would be unchanged. Some restriction is necessary. Here are some possibilities:

1. Drop μ from the model and use I different dummy variables to estimate α_i for $i = 1, \ldots, I$. This is feasible but does not extend well to models with more than one factor as more than one parameter needs to be dropped.

2. Set $\alpha_1 = 0$, then μ represents the expected mean response for the first level and α_i

for $i \neq 1$ represents the difference between level i and level one. Level one is then called the *reference level* or *baseline level*. This can be achieved using treatment contrasts as discussed in the previous chapter.

3. Set $\sum_i \alpha_i = 0$, now μ represents the mean response over all levels and α_i, the difference from that mean. This requires the use of sum contrasts.

The choice of constraint from those listed above or otherwise will determine the coding used to generate the X-matrix. Once that is done, the parameters (effects) can be estimated in the usual way along with standard errors. No matter which valid constraint and coding choice is made, the fitted values and residuals will be the same.

Once the effects are estimated, the natural first step is to test for differences in the levels of the factor. An explicit statement of the null and alternative hypotheses would depend on the coding used. If we use the treatment coding with a reference level, then the null hypothesis would require that $\alpha_2 = \cdots = \alpha_I = 0$. For other codings, the statement would differ. It is simpler to state the hypotheses in terms of models:

$$H_0: \quad y_{ij} \quad = \quad \mu + \varepsilon_{ij}$$
$$H_1: \quad y_{ij} \quad = \quad \mu + \alpha_i + \varepsilon_{ij}$$

We compute the residual sum of squares and degrees of freedom for the two models and then use the same F-test as we have used for regression. The outcome of this test will be the same no matter what coding/restriction we use. If we do not reject the null, we are almost done — we must still check for a possible transformation of the response and outliers. If we reject the null, we might investigate which levels differ.

15.2 An Example

First load the packages:
```
import pandas as pd
import numpy as np
import matplotlib.pyplot as plt
import statsmodels.api as sm
import statsmodels.formula.api as smf
import seaborn as sns
from scipy import stats
import faraway.utils
```
Twenty-four animals were randomly assigned to four different diets and blood samples were taken in a random order. The blood coagulation time was measured. These data come from Box et al. (1978):
```
import faraway.datasets.coagulation
coagulation = faraway.datasets.coagulation.load()
coagulation.head()
   coag diet
1    62    A
2    60    A
3    63    A
4    59    A
5    63    B
```
Some preliminary graphical analysis is essential before fitting. The most popular

plotting method is the side-by-side boxplot. Although this plot has strong visual impact for comparison, plotting the individual points can be better for smaller datasets:
```
sns.boxplot(x="diet", y="coag", data=coagulation)
```
In this data, there are several identical observations. Directly plotting this would obscure this fact. A *swarmplot* perturbs these identical points so that they may be distinguished:
```
sns.swarmplot(x="diet", y="coag", data=coagulation)
```

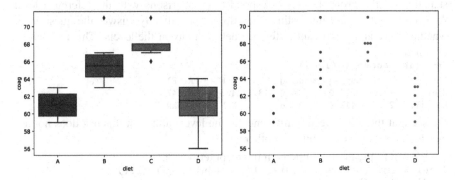

Figure 15.1   A boxplot on the left and a swarmplot on the right showing the blood coagulation data.

The boxplot in Figure 15.1 shows how the four levels vary, but there is something odd about the display of diet C because the median and upper quartile are the same. The swarmplot shows the ties in the data in diets C and D.

We are looking for several features in these plots. We must check for equality of variance in the groups, which seems satisfied in this example. We should also look for evidence of skewness showing a lack of normality. This might suggest a transformation of the response. There is no such concern in this example. Finally, we should look for outliers — there are none to be seen here.

It is not worth considering diagnostics for influence, as the leverages depend explicitly on J_i. If there is only one observation for a given level, that is $J_i = 1$, then the estimate of the effect for that level will be based on that single point. That point is clearly influential without further investigation.

Now let's fit the model using the default treatment coding:
```
lmod = smf.ols("coag ~ diet", coagulation).fit()
lmod.sumary()
          coefs stderr tvalues pvalues
Intercept 61.000 1.183   51.55  0.0000
diet[T.B]  5.000 1.528    3.27  0.0038
diet[T.C]  7.000 1.528    4.58  0.0002
diet[T.D] -0.000 1.449   -0.00  1.0000
```

n=24 p=4 Residual SD=2.366 R-squared=0.67

Diet A is the reference level and has a mean response time of 61, diets B, C and D are 5, 7 and 0 seconds larger, respectively, on average. Examine the design matrix to understand the coding:

```
import patsy
p = patsy.dmatrix('~ diet', coagulation)
p[[0,4,10,16],:]
array([[1., 0., 0., 0.],
       [1., 1., 0., 0.],
       [1., 0., 1., 0.],
       [1., 0., 0., 1.]])
```

We have selected only the first observation of each group to print. The three test statistics for the group levels correspond to comparisons with the reference level A. Although these are interesting, they do not specifically answer the question of whether there is a significant difference between any of the levels. This test can be obtained as:

```
sm.stats.anova_lm(lmod)
```

	df	sum_sq	mean_sq	F	PR(>F)
diet	3.0	228.0	76.0	13.571429	0.000047
Residual	20.0	112.0	5.6	NaN	NaN

We see that there is indeed a difference in the levels although this test does not tell us which levels are different from others.

We can fit the model without an intercept term as in:

```
lmodi = smf.ols("coag ~ diet-1", coagulation).fit()
lmodi.sumary()
```

	coefs	stderr	tvalues	pvalues
diet[A]	61.000	1.183	51.55	0.0000
diet[B]	66.000	0.966	68.32	0.0000
diet[C]	68.000	0.966	70.39	0.0000
diet[D]	61.000	0.837	72.91	0.0000

```
n=24 p=4 Residual SD=2.366 R-squared=0.67
```

We can directly read the level means. To generate the usual test that the means of the levels are equal, we would need to fit the null model and compare using an F-test:

```
lmodnull = smf.ols("coag ~ 1", coagulation).fit()
sm.stats.anova_lm(lmodnull, lmod)
```

	df_resid	ssr	df_diff	ss_diff	F	Pr(>F)
0	23.0	340.0	0.0	NaN	NaN	NaN
1	20.0	112.0	3.0	228.0	13.57	0.000047

We get the same F-statistic and p-value as in the first coding.

We can also use a sum coding:

```
from patsy.contrasts import Sum
lmods = smf.ols("coag ~ C(diet,Sum)", coagulation).fit()
lmods.sumary()
```

	coefs	stderr	tvalues	pvalues
Intercept	64.000	0.498	128.54	0.0000
C(diet, Sum)[S.A]	-3.000	0.974	-3.08	0.0059
C(diet, Sum)[S.B]	2.000	0.845	2.37	0.0282
C(diet, Sum)[S.C]	4.000	0.845	4.73	0.0001

```
n=24 p=4 Residual SD=2.366 R-squared=0.67
```

So the estimated overall mean response is 64 while the estimated mean response for A is three less than the overall mean, that is, 61. Similarly, the means for B and C are 66 and 68, respectively. Since we are using the sum constraint, we compute

$\hat{\alpha}_D = -(-3+2+4) = -3$ so the mean for D is $64-3 = 61$. Notice that $\hat{\sigma}$ and the R^2 are the same as before.

So we can use any of these three methods and obtain essentially the same results. Dropping the intercept is least convenient since an extra step is needed to generate the F-test. Furthermore, the approach would not extend well to experiments with more than one factor, as additional constraints would be needed. The other two methods can be used according to taste. The treatment coding is most appropriate when the reference level is set to a possible control group. We will use the treatment coding by default for the rest of this book.

15.3   Diagnostics

There are fewer diagnostics to do for ANOVA type models, but it is still important to plot the residuals and fitted values and to make the Q–Q plot of the residuals. It makes no sense to transform the predictor, but it is reasonable to consider transforming the response. These diagnostics are shown in Figure 15.2.

```
p = sns.scatterplot(lmod.fittedvalues + \
    np.random.uniform(-0.1,0.1, len(coagulation)), lmod.resid)
p.axhline(0,ls='--')
plt.xlabel("Fitted values")
plt.ylabel("Residuals")
sm.qqplot(lmod.resid, line="q")
```

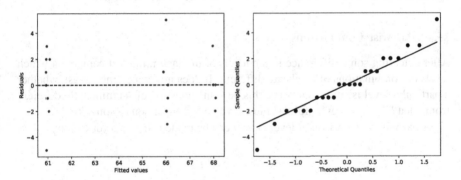

Figure 15.2   Diagnostics for the blood coagulation model.

Because the data are integers and the fitted values happen to be integers also, some discreteness is obvious in the Q–Q plot. Of course, discrete data cannot be normally distributed. However, here the residuals are approximately normal and so we can go ahead with the inference without much concern. The discreteness in the residuals and fitted values shows up in the residual-fitted plot. We have jittered the points so that they can be seen separately.

The assumption of homogeneity of the error variance can be examined using Levene's test. It computes the absolute values of the residuals and uses these as the response in a new one-way ANOVA. To reduce the possible influence of outliers, group medians rather than means should be used. A significant difference would indicate nonconstant variance.

Most tests and confidence intervals (CIs) are relatively insensitive to nonconstant variance so there is no need to take action unless the Levene test is significant at the 1% level.

Applying this to the diet data, we find:

```
coagulation['meds'] = \
    coagulation.groupby('diet').transform(np.median)
coagulation['mads'] = abs(coagulation.coag - coagulation.meds)
lmodb = smf.ols('mads ~ diet', coagulation).fit()
sm.stats.anova_lm(lmodb)
```

	df	sum_sq	mean_sq	F	PR(>F)
diet	3.0	4.333333	1.444444	0.649189	0.592646
Residual	20.0	44.500000	2.225000	NaN	NaN

Since the p-value is large, we conclude that there is no evidence of a nonconstant variance. The `stats.levene` function from `scipy` can be used to achieve the same result using the same syntax as the following test.

An alternative test is due to Bartlett:

```
stats.bartlett(coagulation.coag[coagulation.diet == "A"],
               coagulation.coag[coagulation.diet == "B"],
               coagulation.coag[coagulation.diet == "C"],
               coagulation.coag[coagulation.diet == "D"])
BartlettResult(statistic=1.668, pvalue=0.644)
```

Again, no difference is found. Levene's test is more robust to outliers.

15.4 Pairwise Comparisons

After detecting some difference in the levels of the factor, interest centers on which levels or combinations of levels are different. It does not make sense to ask whether a particular level is significant since this begs the question of "significantly different from what?". Any meaningful test must involve a comparison of some kind.

A pairwise comparison of level i and j can be made using a CI for $\alpha_i - \alpha_j$ using:

$$\hat{\alpha}_i - \hat{\alpha}_j \pm t_{df}^{\alpha/2} se(\hat{\alpha}_i - \hat{\alpha}_j)$$

where $se(\hat{\alpha}_i - \hat{\alpha}_j) = \hat{\sigma}\sqrt{1/J_i + 1/J_j}$ and df $= n - I$ in this case. A test for $\alpha_i = \alpha_j$ amounts to seeing whether zero lies in this interval or not. For example, let's find a 95% CI for $B - A$. From the model output, we can see that the difference is 5.0 with a standard error of 1.53. For differences not involving the reference level of A, more effort would be required to calculate these values. The interval is:

```
lmod.params[1] + \
    np.array([-1, 1]) * stats.t.ppf(0.975,20) * lmod.bse[1]
array([1.814, 8.186])
```

Since zero is not in the interval, the difference is significant. This is fine for just one test, but we are likely to be interested in more than one comparison. Suppose we do all possible pairwise tests when $\alpha = 5\%$ and the null hypothesis is in fact true. In the blood coagulation data, there are four levels and so six possible pairwise comparisons. Even if there was no difference between the four levels, there is still about a 20% chance that at least one significant difference will be found.

For experiments with more levels, the true type I error gets even higher. Using the t-based CIs for multiple comparisons is called the least significant difference (LSD) method, but it can hardly be recommended. Now one might be tempted to argue that we could choose which comparisons are interesting and so reduce the amount of testing and thus the magnitude of the problem. If we only did a few tests, then the Bonferroni adjustment (see Section 6.2.2) could be used to make a simple correction. However, the determination of which comparisons are "interesting" is usually made after seeing the fitted model. This means that all other comparisons are implicitly made even if they are not explicitly computed. On the other hand, if it can be argued that the comparisons were decided before seeing the fit, then we could make the case for the simple adjustment. However, this is rarely the case and furthermore it might be difficult to convince others that this really was your intention. We must usually find a way to adjust for *all* pairwise comparisons.

There are many ways to make the adjustment, but *Tukey's honest significant difference (HSD)* is the easiest to understand. It depends on the studentized range distribution which arises as follows. Let $X_1,\ldots,X_n$ be i.i.d. $N(\mu,\sigma^2)$ and let $R = \max_i X_i - \min_i X_i$ be the range. Then $R/\hat{\sigma}$ has the studentized range distribution $q_{n,\nu}$ where ν is the number of degrees of freedom used in estimating σ.

The Tukey CIs are:

$$\hat{\alpha}_i - \hat{\alpha}_j \pm \frac{q_{l,df}}{\sqrt{2}}\hat{\sigma}\sqrt{(1/J_i + 1/J_j)}$$

When the level sample sizes J_i are very unequal, Tukey's HSD test may become too conservative. We compute the Tukey HSD bands for the B-A difference. The critical value from the studentized range distribution is obtained as:

```
thsd = sm.stats.multicomp.pairwise_tukeyhsd(coagulation.coag,
    coagulation.diet)
thsd.summary()
Multiple Comparison of Means - Tukey HSD,FWER=0.05
===============================================
group1 group2 meandiff  lower    upper   reject
-----------------------------------------------
  A      B      5.0     0.7244   9.2756   True
  A      C      7.0     2.7244  11.2756   True
  A      D      0.0    -4.0562   4.0562  False
  B      C      2.0    -1.8242   5.8242  False
  B      D     -5.0    -8.5773  -1.4227   True
  C      D     -7.0   -10.5773  -3.4227   True
-----------------------------------------------
```

We find that only the $A - D$ and $B - C$ differences are not significant. This merely confirms our impression of which differences are significant. The intervals can be plotted as seen in Figure 15.3.

```
fig = thsd.plot_simultaneous()
```

The Tukey method assumes the worst by focusing on the largest difference. There are other competitors like the Newman–Keuls, Duncan's multiple range and the Waller–Duncan procedure, which are less pessimistic or do not consider all possible pairwise

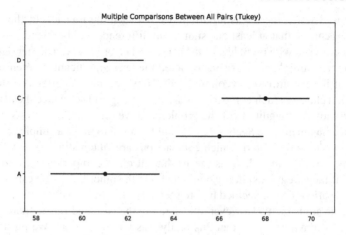

Figure 15.3   Tukey HSD 95% confidence intervals for the pairwise differences.

comparisons. For a detailed description of the many available alternatives, see Hsu (1996).

15.5   False Discovery Rate

Consider some data taken from the Junior School Project collected from primary (US term is elementary) schools in inner London. The data is described in detail in Mortimore et al. (1988). We focus on just two of the variables in the data — the school, of which there are 49, and the mathematics test scores for students from these schools. Suppose we are interested in deviations from the average and so we center the scores:

```
import faraway.datasets.jsp
jsp = faraway.datasets.jsp.load()
jsp['mathcent'] = jsp.math - np.mean(jsp.math)
```

We plot the data as seen in Figure 15.4.

```
sns.boxplot(x="school", y="mathcent", data=jsp)
plt.xticks(fontsize=8, rotation=90)
```

Let's choose the parameterization that omits the intercept term:

```
lmod = smf.ols("mathcent ~ C(school) - 1", jsp).fit()
lmod.sumary()
```

```
               coefs stderr tvalues pvalues
C(school)[1]  -3.368  0.769   -4.38  0.0000
C(school)[2]   0.671  1.229    0.55  0.5848
...edited...
C(school)[50] -2.652  0.734   -3.62  0.0003
```

```
n=3236 p=49 Residual SD=7.372 R-squared=0.08
```

Since we have centered the response, the t-tests, which check for differences from zero, are meaningful. We can see there is good evidence that schools 1 and 50 are significantly below average while the evidence that school 2 is above average is not statistically significant. We can test for a difference between the schools:

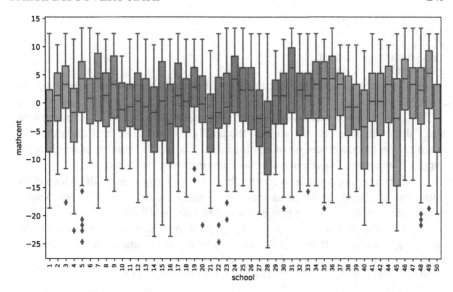

Figure 15.4   Variation-centered math scores by school.

```
sm.stats.anova_lm(smf.ols("mathcent ~ C(school)", jsp).fit())
              df   sum_sq    mean_sq   F       PR(>F)
C(school)     48.0  15483.8   322.57    5.935   1.4580e-33
Residual    3187.0 173212.3    54.35    NaN     NaN
```

We find a strongly significant difference. This comes as little surprise as the sample
size is large, giving us the power to detect quite small differences. Furthermore,
we may have strong prior reasons to expect differences between the schools. A
more interesting question is which schools show clear evidence of under- or over-
performance?

There are too many pairwise comparisons on which to focus our interest. Instead
let us ask which schools have means significantly different from the average. The parameterization we have chosen makes these comparisons easy, but we would expect
about 5% of these differences to be significant even if the null hypothesis held.

Some adjustment is necessary. One approach is to control the *familywise error
rate* (FWER) which is the overall probability of falsely declaring a difference (where
none exists). The *Bonferroni correction* is a simple way to do this. We just multiply
the unadjusted p-values by the number of comparisons. Any probability computed
above one is truncated to one. Let's see which schools have adjusted p-values less
than 5%:

```
from statsmodels.stats.multitest import multipletests
reject, padj, _, _ = multipletests(lmod.pvalues,
    method="bonferroni")
lmod.params[reject]
C(school)[1]    -3.368450
C(school)[16]   -3.737400
C(school)[21]   -3.618450
```

```
C(school)[28]    -5.828595
C(school)[31]     3.824053
C(school)[40]    -4.985458
C(school)[45]    -4.639201
C(school)[50]    -2.652027
```

We see that eight schools are identified with all except school 31 marked as significantly below average.

The Bonferroni correction is known to be conservative, but even were we to use one of the more generous alternatives, the familywise error rate restriction imposes a high bar on the identification of significant effects. As the number of levels being compared increases, this requirement becomes ever more stringent.

An alternative approach is to control the *false discovery rate* (FDR) which is the proportion of effects identified as significant which are not real. The best known method of doing this is due to Benjamini and Hochberg (1995).

Given sorted p-values $p_{(i)}$ for $i = 1, \ldots, m$ the procedure finds the largest i for which $p_{(i)} \leq \alpha i / m$. All tests corresponding to $p_{(i)}$ up to and including this i are declared significant. We compute this for our example:

```
selsch = np.argsort(lmod.pvalues)[np.sort(lmod.pvalues) < \
    np.arange(1,50)*0.05/49]
lmod.params.index[selsch]
Index(['C(school)[28]', 'C(school)[31]', 'C(school)[21]', 'C(school)[1]',
       'C(school)[45]', 'C(school)[40]', 'C(school)[16]', 'C(school)[50]',
       'C(school)[47]', 'C(school)[49]', 'C(school)[4]', 'C(school)[36]',
       'C(school)[46]', 'C(school)[14]', 'C(school)[24]', 'C(school)[27]',
       'C(school)[34]', 'C(school)[9]'],
```

We see that 18 schools are identified compared to the 8 by the previous procedure. FDR is less stringent than FWER in identifying significant effects. A more convenient method of computing the adjusted p-values is:

```
reject, padj, _, _ = multipletests(lmod.pvalues, method="fdr_bh")
lmod.params[reject]
C(school)[1]    -3.368450
C(school)[4]    -2.661928
C(school)[9]     2.245764
...edited...
C(school)[49]    2.796405
C(school)[50]   -2.652027
```

FDR methods are more commonly used where large numbers of comparisons are necessary as often found in imaging or bioinformatics applications. In examples such as these, we expect to find some significant effects and FDR is a useful tool in reliably identifying them.

Exercises

1. In the pulp data, the brightness of the paper produced varies according to four operators.

   (a) Make an appropriate plot of the data. Comment on the content.

   (b) Fit a one-factor model for the bright response. Is there a significant difference between operators?

   (c) Test for a difference in variance in the operators.

(d) If brighter paper is better, can any one operator be considered clearly the best operator? How about the worst operator?

2. In the chickwts data, newly hatched chicks were randomly allocated into six feed supplement groups. Their weights in grams after six weeks is the response. The data can be obtained via the datasets.get_rdataset function in statsmodels.

   (a) Plot the data and comment.

   (b) Fit a model with the weight gain as the response and the feed as a predictor. Is there a difference between the feeds?

   (c) Examine the model summary output. Which feed is the reference level? Which feeds have estimated effects higher than this reference level?

   (d) How many pairwise comparisons are possible for this experiment? Use the Bonferroni correction to compute adjusted p-values from the model output. Which feeds are significantly different from the reference level?

3. The PlantGrowth data shows the weight of plants for three different treatments. The data can be obtained via the datasets.get_rdataset function in statsmodels.

   (a) Make a plot of the data and comment.

   (b) Does the treatment group make a difference to the weight of the plant?

   (c) Use Bartlett's method to test for a difference in variance between the three groups.

   (d) Bartlett's test requires normality of the errors. Is this a reasonable assumption here?

   (e) The control treatment is the existing standard and it would be expensive to change this. Is there enough evidence to switch to one of the other two treatments to achieve a higher weight?

4. Using the infmort data, perform a one-way ANOVA with income as the response and region as the predictor.

   (a) Plot the data and comment on the distributions seen.

   (b) Fit a model to predict income using the region. Is there a significant difference between regions?

   (c) Examine the model summary output. Which region is the reference level? Is there a significant difference between Africa and the Americas?

   (d) Make a multiple pairwise comparison. What regions are different?

   (e) Make a residual-fitted plot and comment.

   (f) Make an appropriate transform on the response and redo the multiple comparisons.

5. The anaesthetic data provides the time to restart breathing unassisted in recovering from general anaesthetic for four treatment groups.

   (a) Produce a boxplot depicting the data. Comment on any features of interest.

   (b) Make a swarmplot of the data. Compare this with the previous plot.

   (c) Modify the previous plot so that tick marks on the y-axis are integers.

(d) Fit a one-factor model for the recovery times and test for a difference between the two groups.

(e) Try a square root transformation on the response. Is the residual-fitted plot satisfactory?

(f) Is there a significant difference among the treatment groups on the transformed scale?

(g) Suppose there is no discernable difference between the four treatments other than a possible difference in recovery time. Which treatment would you choose to reduce recovery time? Can you say what the probability is that you have made the right choice?

6. Data on the butterfat content of milk from Canadian cows of five different breeds can be found in the butterfat dataset. Consider only mature cows.

(a) Filter the data to just mature cows. Plot the data and interpret what you see.

(b) Test for a difference between the breeds.

(c) Make a residual-fitted plot and a qq-plot of the residuals. Do you think your plot in (a) would be sufficient to make the same conclusions?

(d) Compute the leverages and observe. How could these be computed using just the number of parameters and observations?

(e) Produce a plot to check for differences in butterfat between breeds. Which pairwise differences are not statistically significant?

(f) Is there a best breed for producing butterfat? Is there a worst?

7. Five suppliers cut denim for a jeans manufacturer. The amount of waste relative to a target was collected weekly as seen in the denim dataset.

(a) Plot the data to determine which supplier wastes the least. Which supplier is best in terms of minimizing maximum weekly waste?

(b) Is there a significant difference in wastage between the suppliers?

(c) Check the regression diagnostics commenting on any violations.

(d) Remove two outliers and repeat the test for a significant difference. Which supplier has the lowest predicted wastage under this model?

(e) Check for significant pairwise differences between suppliers. Which pairs are significantly different?

(f) Which supplier would you pick if there was no other relevant difference between them? What if the cost of the suppliers was in numerical order with the first the most expensive and the fifth the cheapest?

Chapter 16

Models with Several Factors

In this chapter, we show how to model data with more than one categorical predictor. Sometimes the data can arise from observational studies but such data more commonly arises from designed experiments, often called factorial designs. If all possible combinations of the levels of the factors occur at least once, then we have a full factorial design. Repeated observations for the same combination of factor levels are called *replicates*.

We start with models involving two factors with no replication. It is possible that the factors can interact, but this is difficult to investigate without replication. We consider examples with and without significant interaction and discuss how they should be interpreted. Replication can be expensive, so sometimes it is better to use the experimental resources to investigate more factors. This leads us to an example with many factors but no replication.

16.1 Two Factors with No Replication

Load the packages:
```
import pandas as pd
import numpy as np
import matplotlib.pyplot as plt
import statsmodels.api as sm
import statsmodels.formula.api as smf
import seaborn as sns
from scipy import stats
import faraway.utils
```
Mazumdar and Hoa (1995) report an experiment to test the strength of a thermoplastic composite depending on the power of a laser and the speed of a tape:
```
import faraway.datasets.composite
composite = faraway.datasets.composite.load()
composite
```
	strength	laser	tape
1	25.66	40W	slow
2	29.15	50W	slow
3	35.73	60W	slow
4	28.00	40W	medium
5	35.09	50W	medium
6	39.56	60W	medium
7	20.65	40W	fast
8	29.79	50W	fast
9	35.66	60W	fast

We can plot the data with seaborn to produce a color display (not shown)
```
sns.catplot(x="laser", y="strength", hue="tape",
```

```
   data=composite, kind="point")
```

Alternatively, we can plot using `matplotlib` which gives us more control over the appearance at the cost of some complexity. We find it helpful to create a numeric version of the `laser` variable.

```
composite['Nlaser'] = np.tile([40,50,60],3)
faclevels = np.unique(composite.tape)
lineseq = ['-','--',":"]
for i in np.arange(len(faclevels)):
    j = (composite.tape == faclevels[i])
    plt.plot(composite.Nlaser[j], composite.strength[j],
             lineseq[i],label=faclevels[i])
plt.legend()
plt.xlabel("Laser")
plt.ylabel("Strength")
plt.xticks([40,50,60],["40W","50W","60W"])
```

We can reverse the roles of `laser` and `tape` as in this seaborn plot (not shown):

```
sns.catplot(x="tape", y="strength", hue="laser",
            data=composite, kind="point")
```

We can also plot it using `matplotlib`. We use the numerical values for the tape speed.

```
composite['Ntape'] = np.repeat([6.42,13,27], 3)
faclevels = np.unique(composite.laser)
for i in np.arange(len(faclevels)):
    j = (composite.laser == faclevels[i])
    plt.plot(composite.Ntape[j], composite.strength[j],
             lineseq[i],label=faclevels[i])
plt.legend()
plt.xlabel("Tape")
plt.ylabel("Strength")
plt.xticks([6.42,13,27],["slow","medium","fast"])
```

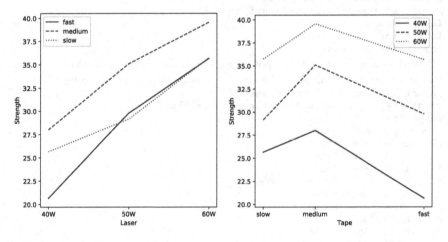

Figure 16.1   Plots for the composite data — the role of the predictors is reversed in each plot.

The data has been plotted in Figure 16.1 with roles of the two factors shown in both ways to give better insight.

We can see in the left plot that medium tape speed is best for strength. while in the right we see that the 60W setting for laser is best. When there is little or no interaction effect, we would expect the lines in each plot to be approximately parallel (i.e., a constant distance apart). This would be a sign that the effects are additive. Of course, we must allow for a certain amount of noise clouding the observation, but in this example, the lines do appear to be roughly parallel.

A general model for this type of data is:

$$y_{ij} = \mu + \alpha_i + \beta_j + (\alpha\beta)_{ij} + \varepsilon_{ij}$$

where $i = 1, \ldots I$ and $j = 1, \ldots J$. The $\alpha\beta$ terms represent interactions between the levels of α and β.

We fit this model to the current example:

```
lmod = smf.ols("strength ~ tape * laser", composite).fit()
lmod.sumary()
```

	coefs	stderr	tvalues	pvalues
Intercept	20.650	inf	0.00	nan
tape[T.medium]	7.350	inf	0.00	nan
tape[T.slow]	5.010	inf	0.00	nan
laser[T.50W]	9.140	inf	0.00	nan
laser[T.60W]	15.010	inf	0.00	nan
tape[T.medium]:laser[T.50W]	-2.050	inf	-0.00	nan
tape[T.slow]:laser[T.50W]	-5.650	inf	-0.00	nan
tape[T.medium]:laser[T.60W]	-3.450	inf	-0.00	nan
tape[T.slow]:laser[T.60W]	-4.940	inf	-0.00	nan

```
n=9 p=9 Residual SD=inf R-squared=1.00
```

We have parameter estimates but nothing else. The model has as many parameters as cases, so a perfect fit has been achieved. Unfortunately, no further inference is possible and little has been gained from this model.

We cannot check for an interaction directly by adding such a term to the model, but Tukey's nonadditivity test provides another way of investigating for an interaction. The model:

$$y_{ij} = \mu + \alpha_i + \beta_j + \phi\alpha_i\beta_j + \varepsilon_{ijk}$$

is fit to the data and then we test if $\phi = 0$. This is a nonlinear model because it involves a product of parameters, $\phi\alpha_i\beta_j$ which is difficult to fit in general. We get around this problem by fitting the model in stages. First we fit the model with no interaction and extract the estimates of the main effects.

```
lmod = smf.ols("strength ~ tape + laser", composite).fit()
lmod.params.index
Index(['Intercept', 'tape[T.medium]', 'tape[T.slow]', 'laser[T.50W]',
        'laser[T.60W]'])
```

Since the tape effects are not in the expected order, we need to construct these carefully:

```
tapecoefs = np.repeat([lmod.params[2],lmod.params[1],0], 3)
```

and similarly for the laser effects:

```
lasercoefs = np.tile(np.append(0, lmod.params[3:5]),3)
```

256   MODELS WITH SEVERAL FACTORS

Now we update the model with the new predictor formed from the products of the
main effects:
```
composite['crossp'] = tapecoefs * lasercoefs
tmod = smf.ols("strength ~ tape + laser + crossp",
    composite).fit()
tmod.sumary()
```
```
                 coefs stderr tvalues pvalues
Intercept       21.948  1.491   14.72  0.0007
tape[T.medium]   6.675  2.239    2.98  0.0585
tape[T.slow]     1.791  1.498    1.20  0.3178
laser[T.50W]     7.087  1.618    4.38  0.0220
laser[T.60W]    13.168  2.014    6.54  0.0073
crossp          -0.034  0.050   -0.67  0.5503
```
```
n=9 p=6 Residual SD=1.745 R-squared=0.97
```
The p-value of 0.5503 indicates a nonsignificant interaction. So for these data, we
might reasonably assume $(\alpha\beta)_{ij} = 0$. The major drawback is that the test makes the
assumption that the interaction effect is multiplicative in form. We have no particular
reason to believe it takes this form, and so this alternative hypothesis may not be
looking in the right place. Although judging whether lines on a plot are parallel
requires some subjective judgment, it may be a more reliable method of checking for
interaction here.

Now that the issue of interactions has been addressed, we can check the significance of the main effects:
```
sm.stats.anova_lm(tmod)
```
```
           df       sum_sq       mean_sq           F    PR(>F)
tape      2.0    48.918689     24.459344    8.034443  0.062401
laser     2.0   224.183822    112.091911   36.820124  0.007745
crossp    1.0     1.370111      1.370111    0.450056  0.550338
Residual  3.0     9.132933      3.044311         NaN       NaN
```
We see that both factors are significant.

The treatment coding does not take advantage of the ordered nature of both factors. Factors without an ordering to the levels are called *nominal*, while those that
possess a natural ordering are called *ordinal*. We can declare both to be *ordered
factors* and refit:
```
cat_type = pd.api.types.CategoricalDtype(
    categories=['slow','medium','fast'],ordered=True)
composite['tape'] = composite.tape.astype(cat_type)
from patsy.contrasts import Poly
lmod = smf.ols("strength ~ C(tape,Poly) + C(laser,Poly)",
    composite).fit()
lmod.sumary()
```
```
                        coefs stderr tvalues pvalues
Intercept              31.032  0.540   57.45  0.0000
C(tape, Poly).Linear   -1.047  0.936   -1.12  0.3259
C(tape, Poly).Quadratic -3.900  0.936   -4.17  0.0140
C(laser, Poly).Linear   8.636  0.936    9.23  0.0008
C(laser, Poly).Quadratic -0.381  0.936  -0.41  0.7047
```
```
n=9 p=5 Residual SD=1.620 R-squared=0.96
```
Instead of a coding with respect to a reference level, we have linear and quadratic
terms for each factor. The coding is:

```
from patsy import dmatrix
dm = dmatrix('~ C(tape,Poly) + C(laser,Poly)', composite)
np.asarray(dm).round(2)
array([[ 1.  , -0.71,  0.41, -0.71,  0.41],
       [ 1.  , -0.71,  0.41, -0.  , -0.82],
       [ 1.  , -0.71,  0.41,  0.71,  0.41],
       [ 1.  , -0.  , -0.82, -0.71,  0.41],
       [ 1.  , -0.  , -0.82, -0.  , -0.82],
       [ 1.  , -0.  , -0.82,  0.71,  0.41],
       [ 1.  ,  0.71,  0.41, -0.71,  0.41],
       [ 1.  ,  0.71,  0.41, -0.  , -0.82],
       [ 1.  ,  0.71,  0.41,  0.71,  0.41]]])
```

We see the linear term is proportional to $(-1, 0, 1)$ representing a linear trend across the levels, while the quadratic term is proportional to $(1, -2, 1)$ representing a quadratic trend.

We see that the quadratic term for laser power is not significant, while there is a quadratic effect for tape speed. One of the drawbacks of a model with factors is the difficulty of extrapolating to new conditions. The information gained from the ordered factors suggests a model with numerical predictors corresponding to the level values.

```
lmodn = smf.ols(
    "strength ~ np.log(Ntape) + I(np.log(Ntape)**2) + Nlaser",
    composite).fit()
lmodn.sumary()
```

	coefs	stderr	tvalues	pvalues
Intercept	-55.050	13.338	-4.13	0.0091
np.log(Ntape)	46.593	10.499	4.44	0.0068
I(np.log(Ntape) ** 2)	-9.238	2.028	-4.55	0.0061
Nlaser	0.611	0.060	10.11	0.0002

n=9 p=4 Residual SD=1.479 R-squared=0.96

We use the log of tape speed, as this results in roughly evenly spaced levels. This model fits about as well as the two-factor model, but has the advantage that we make predictions for values of tape speed and laser power that were not used in the experiment. The earlier analysis with factors alone helped us discover this model, which we may not otherwise have found.

16.2 Two Factors with Replication

Consider the case when the number of observations per combination of factor levels is the same and greater than one. Such a layout results in an orthogonal design matrix. With the benefit of replication, we are now free to fit and test the full model:

$$y_{ijk} = \mu + \alpha_i + \beta_j + (\alpha\beta)_{ij} + \varepsilon_{ijk}$$

The interaction effect is tested by fitting a model without the $(\alpha\beta)_{ij}$ term and computing the usual F-test. If the interaction effect is found to be significant, we do not test the main effects even if they appear not to be significant.

In an experiment to study factors affecting the production of the polyvinyl chloride (PVC) plastic, three operators used eight different devices called resin railcars to produce PVC. For each of the 24 combinations, two samples were produced. The response is the particle size of the product. The experiment is described in Morris and Watson (1998).

We make plots with respect to each of the predictors, as seen in Figure 16.2:

```
import faraway.datasets.pvc
pvc = faraway.datasets.pvc.load()
```

First using seaborn (not shown).

```
sns.catplot(x='resin',y='psize',hue='operator', data=pvc,
    kind="point",ci=None)
```

Producing the plot in matplotlib requires us to first average over the two replications:

```
pvcm = pvc.groupby(
    ['operator','resin'])[['psize']].mean().reset_index()
```

Now we can construct the plot:

```
faclevels = np.unique(pvcm.operator)
lineseq = ['-','--',":"]
for i in np.arange(len(faclevels)):
    j = (pvcm.operator == faclevels[i])
    plt.plot(pvcm.resin[j], pvcm.psize[j],
             lineseq[i],label=faclevels[i])
plt.legend(title="Operator")
plt.xlabel("resin")
plt.ylabel("Particle Size")
```

We can reverse the roles of the factors with:

```
sns.catplot(x='operator',y='psize',hue='resin', data=pvc,
    ci=None,scale=0.5,kind="point")
```

or produce in matplotlib.

```
faclevels = np.unique(pvcm.resin)
for i in np.arange(len(faclevels)):
    j = (pvcm.resin == faclevels[i])
    plt.plot(pvcm.operator[j], pvcm.psize[j])
    plt.annotate(str(i+1), [0.95,np.ravel(pvcm.psize[j])[0]])
plt.xlabel("Operator")
plt.xticks([1,2,3])
plt.ylabel("Particle Size")
```

The eight resins are difficult to distinguish using eight linetypes, so we have labeled the line with a number at the left.

In the left plot, there are eight levels of resin. For each level of resin, we observe two responses from each of the three operators which we have averaged. We have plotted the operators each with a different line type. The variances for each level of resin seem approximately equal. The three lines are approximately parallel, suggesting little interaction.

In the right plot, we have drawn lines connecting the mean values for each level of the resin. We can see from this plot that particle sizes tend to be largest for the first operator and smallest for the third. We see that the variance of the response is about the same for each operator. We have annotated the plot to label the eight levels so that they can be readily distinguished. Note that it is difficult to distinguish eight different line styles so we have not attempted this. We can see that the lines are approximately parallel which leads us to expect that there is little interaction between the two factors.

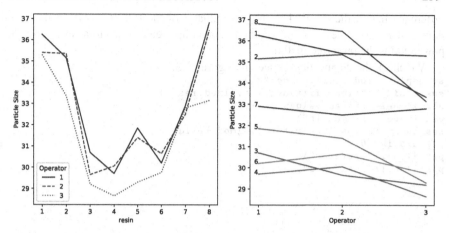

Figure 16.2   PVC data plotted in two ways. Lines connect the average responses for the combination of levels. The railcar number appears to the left of each line on the right plot.

The trouble with interaction plots is that we always expect there to be some random variation, so it is sometimes difficult to distinguish true interaction from just noise. Fortunately, in this case, we have replication so we can directly test for an interaction effect. We fit the full model and check the significance of the factors. We need to convert the resin and operator to categorical variables as previously they have been numeric. If we did not make this conversion, they would be fitted as numeric predictors.

```
pvc['operator'] = pvc['operator'].astype('category')
pvc['resin'] = pvc['resin'].astype('category')
lmod = smf.ols('psize ~ operator*resin', pvc).fit()
sm.stats.anova_lm(lmod).round(3)
```

	df	sum_sq	mean_sq	F	PR(>F)
operator	2.0	20.718	10.359	7.007	0.004
resin	7.0	283.946	40.564	27.439	0.000
operator:resin	14.0	14.335	1.024	0.693	0.760
Residual	24.0	35.480	1.478	NaN	NaN

We see that the interaction effect is not significant. This now allows a meaningful investigation of the significance of the main effects. We see that both main effects are significant.

Now one may be tempted to remove the interaction and then retest the main effects:

```
lmod2 = smf.ols('psize ~ operator+resin', pvc).fit()
sm.stats.anova_lm(lmod2).round(3)
```

	df	sum_sq	mean_sq	F	PR(>F)
operator	2.0	20.718	10.359	7.902	0.001
resin	7.0	283.946	40.564	30.943	0.000
Residual	38.0	49.815	1.311	NaN	NaN

One gets a somewhat different result although the conclusions are the same. The difference lies in the $\hat{\sigma}^2$ used in the denominator of the F-tests (1.478 in the first table and 1.311 in the second). The first version has been shown to achieve greater power and is the preferred method.

We check the diagnostics, as seen in Figure 16.3:

```
sm.qqplot(lmod.resid, line="q")
sns.residplot(lmod.fittedvalues, lmod.resid)
plt.xlabel("Fitted values")
plt.ylabel("Residuals")
sns.swarmplot(pvc['operator'], lmod.resid)
plt.axhline(0)
plt.xlabel("Operator")
plt.ylabel("Residuals")
```

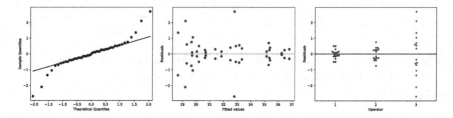

Figure 16.3   Diagnostics plots for the full model for the PVC data.

The Q–Q plot suggests that the errors are somewhat long-tailed. We can see some symmetry in the residuals vs. fitted plot. For each combination of the factors, the mean of the two replicates is the fitted value. The two residuals for that pair will be mirror images. A closer look at the more distant pairs reveals that they derive from operator 3 as seen in the right panel of Figure 16.3.

Although the plot is fairly convincing, we can test for a difference in the variance. For each pair of observations coming from a particular combination of operator and resin, we compute the two residuals. One will be positive and the other negative due to the symmetry previously noted. We take $\sqrt{|\varepsilon|}$ as the new response. This measures the spread in the pair with the square root transformation reducing the skewness in the distribution. Note that we only need one observation from each pair since these are identical.

```
lmod = smf.ols('psize ~ operator*resin', pvc).fit()
ii = np.arange(0,48,2)
pvce = pvc.iloc[ii,:].copy()
pvce['res'] = np.sqrt(abs(lmod.resid.loc[ii+1]))
vmod = smf.ols('res ~ operator+resin', pvce).fit()
sm.stats.anova_lm(vmod).round(3)
```

	df	sum_sq	mean_sq	F	PR(>F)
operator	2.0	1.490	0.745	15.117	0.000
resin	7.0	0.638	0.091	1.848	0.155
Residual	14.0	0.690	0.049	NaN	NaN

This confirms a significant difference in variation for the operators. It also tells us there is no strong evidence against constant variation among the resin cars.

This may be the most important finding from such an experiment because consistency in manufacturing is very important. In many situations, one wishes to manufacture an object to a specification. It is often possible to adjust the mean to the desired level but a high variance will be detrimental to production quality.

Even so, let us examine the main effects, bearing in mind that we do not have constant variance of the errors, making the comparisons less than optimal.

```
lmod = smf.ols('psize ~ operator+resin', pvc).fit()
lmod.sumary()
            coefs stderr tvalues pvalues
Intercept   36.240 0.523  69.34  0.0000
operator[T.2] -0.263 0.405  -0.65  0.5206
operator[T.3] -1.506 0.405  -3.72  0.0006
resin[T.2]   -1.033 0.661  -1.56  0.1263
resin[T.3]   -5.800 0.661  -8.77  0.0000
resin[T.4]   -6.183 0.661  -9.35  0.0000
resin[T.5]   -4.800 0.661  -7.26  0.0000
resin[T.6]   -5.450 0.661  -8.24  0.0000
resin[T.7]   -2.917 0.661  -4.41  0.0001
resin[T.8]   -0.183 0.661  -0.28  0.7830
```

n=48 p=10 Residual SD=1.145 R-squared=0.86

Suppose that we valued a small particle size. In this case, operator 3 produces the best results overall. If we have a choice, we would also prefer resin car 4 for the smallest particle size. However, sometimes we may not have control over a factor. For example, the resin cars might be disposable and we cannot reliably produce one to a particular specification. In such cases, there is no point in identifying which is best in the experiment. Nevertheless, we would still include the factor in the model because it would allow us to estimate the operator effects more precisely.

We can construct pairwise confidence intervals for the treatment factors using the Tukey method. First we construct the width of the interval for the operator differences:

```
from statsmodels.sandbox.stats.multicomp import get_tukeyQcrit
get_tukeyQcrit(3,38) * lmod.bse[1] / np.sqrt(2)
```

We can compute all the pairwise differences among operators with:

```
p = np.append(0,lmod.params[1:3])
np.add.outer(p,-p)
array([[ 0.     , 0.2625 , 1.50625],
       [-0.2625 , 0.     , 1.24375],
       [-1.50625, -1.24375, 0.     ]])
```

We see that operators 1 and 2 are not significantly different, but operator 3 is different from both. There are more significant differences among the resin cars although this may not be of specific interest.

The analysis above is appropriate for the investigation of specific operators and resin cars. These factors are being treated as *fixed effects*. If the operators and resin cars were randomly selected from larger populations of those available, they should be analyzed as *random effects*. This would require a somewhat different analysis not covered here. However, we can at least see from the analysis above that the variation between resin cars is greater than that between operators.

It is important that the observations taken in each cell are genuine replications. If this is not true, then the observations will be correlated and the analysis will need to be adjusted. It is a common scientific practice to repeat measurements and take the average to reduce measurement errors. These repeat measurements are not independent observations. Data where the replicates are correlated can be handled with repeated measures models. For example, in this experiment we would need to take some care to separate the two measurements for each operator and resin car. Some knowledge of the engineering might be necessary to achieve this.

16.3 Two Factors with an Interaction

An industrial experiment examined the number of warp breaks in yarn depending on two types of wool and three levels of tension used in a weaving machine. The data may be found in Tukey (1977):

```
import faraway.datasets.warpbreaks
warpbreaks = faraway.datasets.warpbreaks.load()
warpbreaks.head()
   breaks wool tension
1      26    A       L
2      30    A       L
3      54    A       L
4      25    A       L
5      70    A       L
```

We could make a simple boxplot (not shown):

```
sns.boxplot(x="wool", y="breaks", data=warpbreaks)
```

We could make a fancier plot (not shown) with seaborn using:

```
ax = sns.catplot(x='wool',y='breaks',hue='tension',
    data=warpbreaks, ci=None,kind="point")
ax = sns.swarmplot(x='wool',y='breaks',hue='tension',
    data=warpbreaks)
ax.legend_.remove()
```

We can also make plots using matplotlib. In the right plot of Figure 16.4, we create a numeric version of wool and use this to offset the levels of wool so that they can be easily distinguished:

```
warpbreaks['nwool'] = np.where(warpbreaks['wool'] == 'A',0,1)
faclevels = np.unique(warpbreaks.tension)
markseq = [".","x","+"]
for i in np.arange(len(faclevels)):
    j = (warpbreaks.tension == faclevels[i])
    plt.scatter(warpbreaks.nwool[j]+0.05*i, warpbreaks.breaks[j],
                marker=markseq[i],label=faclevels[i])
plt.legend(title="Tension")
plt.xlabel("Wool")
plt.xticks([0,1],["A","B"])
plt.ylabel("Warpbreaks")
```

It is also helpful to plot the means within each combination of wool and tension as seen in the left panel of Figure 16.4.

```
wm = warpbreaks.groupby(
    ['wool','tension'])[['breaks']].mean().reset_index()
faclevels = np.unique(wm.tension)
lineseq = ['-','--',":"]
for i in np.arange(len(faclevels)):
    j = (wm.tension == faclevels[i])
    plt.plot(wm.wool[j], wm.breaks[j],lineseq[i],
             label=faclevels[i])
plt.legend(title="Tension")
plt.xlabel("Wool")
```

```
plt.ylabel("Warpbreaks")
```

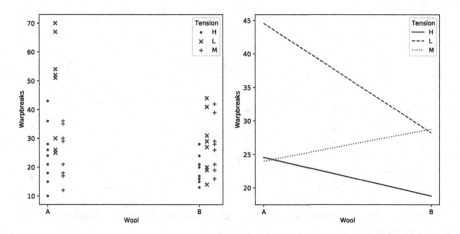

Figure 16.4   Plots showing how the number of warpbreaks vary by wool type for each level of tension.

The plots should be repeated, as seen in Figure 16.5, reversing the roles of the two predictors.

```
warpbreaks['ntension'] = \
    np.where(warpbreaks['tension'] == 'L',
    0,np.where(warpbreaks['tension']=='M',1,2))
faclevels = np.unique(warpbreaks.wool)
markseq = ["x","+"]
for i in np.arange(len(faclevels)):
    j = (warpbreaks.wool == faclevels[i])
    plt.scatter(warpbreaks.ntension[j]+0.1*i,
                warpbreaks.breaks[j],
                marker=markseq[i],
                label=faclevels[i])
plt.legend(title="Wool")
plt.xlabel("Tension")
plt.xticks([0,1,2],["L","M","H"])
plt.ylabel("Warpbreaks")
```

and similarly for the means:

```
lineseq = ["-",":"]
for i in np.arange(len(faclevels)):
    j = (wm.wool == faclevels[i])
    plt.plot(wm.tension[j], wm.breaks[j],
            lineseq[i],label=faclevels[i])
plt.legend(title="Wool")
plt.xlabel("Tension")
plt.ylabel("Warpbreaks")
```

There is some evidence of nonconstant variation. We can also see there may be some interaction between the factors, as the lines joining the mean response at each level are not close to parallel.

We can now fit a model with an interaction effect and check the diagnostics:

```
lmod = smf.ols('breaks ~ wool * tension', warpbreaks).fit()
```

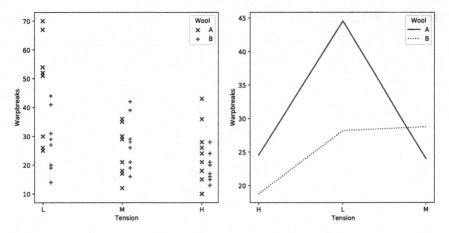

Figure 16.5   Plots showing how the number of warpbreaks vary by tension level for each type of wool.

```
fig, ax = plt.subplots()
sns.regplot(lmod.fittedvalues, lmod.resid,
    scatter_kws={'alpha':0.3}, ci=None, ax=ax)
ax.set_xlim(np.array(ax.get_xlim())+[-1,1])
ax.set_xlabel("Fitted values")
ax.set_ylabel("Residuals")
```

The residuals vs. fitted plot shown in the right panel of Figure 16.6 reveals increasing variation with the response. Given that the response is a counted variable, a square root transformation is the suggested solution:

```
lmod = smf.ols('np.sqrt(breaks) ~ wool * tension',
    warpbreaks).fit()
fig, ax = plt.subplots()
sns.regplot(lmod.fittedvalues, lmod.resid,
    scatter_kws={'alpha':0.3}, ci=None, ax=ax)
ax.set_xlim(np.array(ax.get_xlim())+[-0.1,0.1])
ax.set_xlabel("Fitted values")
ax.set_ylabel("Residuals")
```

The diagnostic plot shown in the left panel of Figure 16.6 shows an improvement. We can now test for the significance of the effects:

```
sm.stats.anova_lm(lmod).round(4)
```

	df	sum_sq	mean_sq	F	PR(>F)
wool	1.0	2.9019	2.9019	3.0222	0.0885
tension	2.0	15.8916	7.9458	8.2752	0.0008
wool:tension	2.0	7.2014	3.6007	3.7500	0.0307
Residual	48.0	46.0892	0.9602	NaN	NaN

We see that there is a significant interaction effect between the factors. This means we cannot express the effect of the choice of wool independently of the setting of the tension. We must interpret their effects together. One may also notice that the main effect for wool is not significant but this does not mean that the choice of wool has no effect. To make this point clearer, consider the regression output:

```
lmod.sumary()
```

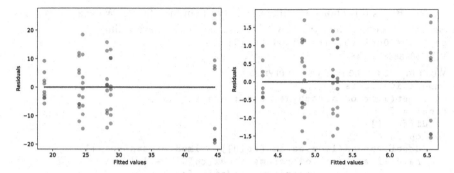

Figure 16.6   Diagnostic plots for the warpbreaks data. The plot on the left comes from the untransformed response while that on the right results from a square-rooted response.

	coefs	stderr	tvalues	pvalues
Intercept	4.856	0.327	14.87	0.0000
wool[T.B]	-0.554	0.462	-1.20	0.2361
tension[T.L]	1.691	0.462	3.66	0.0006
tension[T.M]	-0.030	0.462	-0.07	0.9477
wool[T.B]:tension[T.L]	-0.755	0.653	-1.16	0.2534
wool[T.B]:tension[T.M]	1.027	0.653	1.57	0.1225

n=54 p=6 Residual SD=0.980 R-squared=0.36

In this output, we see that the main effect term for wool is significant. The reason for the discrepancy is that the two tests of wool are different. The former test compares the wool choice averaged over all levels of tension while the latter makes the comparison at the reference level of tension (low in this case). So there is no simple answer to the question of whether the choice of wool makes a difference to the response. In the presence of a significant interaction, it is better not to consider main effects alone but consider the effect of the two factors together.

The regression output above does not make it easy to interpret the nature of the interaction. Sometimes it can be easier to think of the model having just one combined factor with a number of levels equal to the product of the two factor levels. We can achieve this as follows:

```
lmod = smf.ols(
    'np.sqrt(breaks) ~ wool:tension-1', warpbreaks).fit()
lmod.sumary()
```

	coefs	stderr	tvalues	pvalues
wool[A]:tension[H]	4.856	0.327	14.87	0.0000
wool[B]:tension[H]	4.302	0.327	13.17	0.0000
wool[A]:tension[L]	6.548	0.327	20.05	0.0000
wool[B]:tension[L]	5.238	0.327	16.04	0.0000
wool[A]:tension[M]	4.826	0.327	14.77	0.0000
wool[B]:tension[M]	5.299	0.327	16.22	0.0000

n=54 p=6 Residual SD=0.980 R-squared=0.36

We can see that wool B at high tension produces the lowest expected number of breaks. But is this choice significantly better than the others? We can answer this question by computing the Tukey HSD intervals which have width:

```
get_tukeyQcrit(6,48) * lmod.bse[0]
1.3725048838137048
```

We compute all the pairwise intervals:

```
import itertools
dp = set(itertools.combinations(range(0,6),2))
dcoef = []
namdiff = []
for cp in dp:
    dcoef.append(lmod.params[cp[0]] - lmod.params[cp[1]])
    namdiff.append(lmod.params.index[cp[0]] + '-' + \
                   lmod.params.index[cp[1]])
thsd = pd.DataFrame({'Difference':dcoef},index=namdiff)
thsd["lb"] = thsd.Difference - get_tukeyQcrit(6,48) * lmod.bse[0]
thsd["ub"] = thsd.Difference + get_tukeyQcrit(6,48) * lmod.bse[0]
thsd.round(3)
```

	Difference	lb	ub
wool[A]:tension[H]-wool[B]:tension[H]	0.554	-0.818	1.927
wool[B]:tension[H]-wool[A]:tension[L]	-2.245	-3.618	-0.873
wool[B]:tension[H]-wool[B]:tension[L]	-0.936	-2.308	0.437
wool[A]:tension[M]-wool[B]:tension[M]	-0.473	-1.845	0.900
wool[B]:tension[H]-wool[A]:tension[M]	-0.524	-1.896	0.849
wool[B]:tension[H]-wool[B]:tension[M]	-0.996	-2.369	0.376
wool[A]:tension[L]-wool[A]:tension[M]	1.722	0.349	3.094
wool[A]:tension[H]-wool[B]:tension[M]	-0.442	-1.815	0.930
wool[A]:tension[L]-wool[B]:tension[L]	1.309	-0.063	2.682
wool[A]:tension[L]-wool[B]:tension[M]	1.249	-0.124	2.621
wool[A]:tension[H]-wool[A]:tension[M]	0.030	-1.342	1.403
wool[A]:tension[H]-wool[B]:tension[L]	-0.382	-1.754	0.991
wool[B]:tension[L]-wool[A]:tension[M]	0.412	-0.960	1.785
wool[A]:tension[H]-wool[A]:tension[L]	-1.691	-3.064	-0.319
wool[B]:tension[L]-wool[B]:tension[M]	-0.061	-1.433	1.312

We see that no pairwise difference involving medium or high tensions is significant. So the advice should be to avoid running the machine at low tension if possible. If we must make a choice, using wool B at high tension is most likely to be the best choice. But if there are cost differences between A and B or difficulties in running the machines at high tension, then we can see that other choices may be reasonable.

16.4 Larger Factorial Experiments

Suppose we have factors $\alpha, \beta, \gamma, \ldots$ at levels $l_\alpha, l_\beta, l_\gamma, \ldots$. A *full* factorial experiment has at least one run for each combination of the levels. The number of combinations is $l_\alpha l_\beta l_\gamma \ldots$, which could easily be very large. The biggest model for a full factorial contains all possible interaction terms, which range from second-order, or two-way, as encountered earlier in this chapter, to high order interactions involving several factors. For this reason, full factorials are rarely executed for more than three or four factors.

There are some advantages to factorial designs. If no interactions are significant, we get several one-way experiments for the price of one. Compare this with doing a sequence of one-way experiments. It is sometimes better to use replication for investigating another factor instead. For example, instead of doing a two-factor experiment with replication, we might use that replication to investigate a third factor but now with no replication.

The analysis of full factorial experiments is an extension of that used for the two-way ANOVA. Typically, there is no replication due to cost concerns, so it is necessary to assume that some higher order interactions are zero in order to free up degrees of freedom for testing the lower order effects. Not many phenomena require a precise combination of several factors so this is not unreasonable.

Fractional factorials

Fractional factorials use only a fraction of the number of runs in a full factorial experiment. This is done to save the cost of the full experiment or to make only a few runs because the experimental material is limited. It is often possible to estimate the lower order effects with just a fraction. Consider an experiment with seven factors, each at two levels:

Effect	mean	main	2-way	3-way	4	5	6	7
Number of parameters	1	7	21	35	35	21	7	1

Table 16.1 Number of parameters in a two-level, seven-factor experiment.

If we are going to assume that higher order interactions are negligible, then we do not really need $2^7 = 128$ runs to estimate the remaining parameters. We could perform only eight runs and still be able to estimate the seven main effects, though none of the interactions. In this particular example, it is hard to find a design to estimate all the two-way interactions uniquely, without a large number of runs. The construction of good designs is a complex task. For example, see Hamada and Wu (2000) for more on this. A Latin square (see Section 17.2) where all predictors are considered as factors is another example of a fractional factorial.

In fractional factorial experiments, we try to estimate many parameters with as few data points as possible. This means there are often not many degrees of freedom left. We require that σ^2 be small; otherwise there will be little chance of distinguishing significant effects. Fractional factorials are popular in engineering applications where the experiment and materials can be tightly controlled. Fractional factorials are commonly found in product design because they allow for the screening of a large number of factors. Factors identified in a screening experiment can then be more closely investigated. In the social sciences and medicine, the experimental materials, often human or animal, are much less homogeneous and less controllable, so σ^2 tends to be relatively large. In such cases, fractional factorials are of no value.

Here is an example. Speedometer cables can be noisy because of shrinkage in the plastic casing material. An experiment was conducted to find out what caused shrinkage by screening a large number of factors. The engineers started with 15 different factors: liner outside diameter, liner die, liner material, liner line speed, wire braid type, braiding tension, wire diameter, liner tension, liner temperature, coating material, coating die type, melt temperature, screen pack, cooling method and line speed, labeled a through o. Response is percentage of shrinkage per specimen. There were two levels of each factor. The "+" indicates the high level of a factor and the "−" indicates the low level.

A full factorial would take 2^{15} runs, which is highly impractical, thus a design with only 16 runs was used where the particular runs have been chosen specially so as to estimate the mean and the 15 main effects. We assume that there is no interaction effect of any kind. The data come from Box et al. (1988).

Here is the data:
```
import faraway.datasets.speedo
speedo = faraway.datasets.speedo.load()
speedo
```

	h	d	l	b	j	f	n	a	i	e	m	c	k	g	o	y
1	-	-	+	-	+	+	-	-	+	+	-	+	-	-	+	0.4850
2	+	-	-	-	-	+	+	-	-	+	+	+	+	-	-	0.5750
3	-	+	-	-	+	-	+	-	+	-	+	-	+	-	+	0.0875
4	+	+	+	-	-	-	-	-	-	-	-	+	+	+	+	0.1750
5	-	-	+	+	-	-	+	-	+	+	-	-	+	+	-	0.1950
6	+	-	-	+	+	-	-	-	-	+	+	-	-	+	+	0.1450
7	-	+	-	+	-	+	-	-	+	-	+	-	+	-	+	0.2250
8	+	+	+	+	+	+	+	-	-	-	-	-	-	-	-	0.1750
9	-	-	+	-	+	+	-	+	-	-	+	-	+	+	-	0.1250
10	+	-	-	-	-	+	+	+	+	-	-	-	-	+	+	0.1200
11	-	+	-	-	+	-	+	+	-	+	-	-	+	-	+	0.4550
12	+	+	+	-	-	-	-	+	+	+	+	-	-	-	-	0.5350
13	-	-	+	+	-	-	+	+	-	-	+	+	-	-	+	0.1700
14	+	-	-	+	+	-	-	+	+	-	-	+	+	-	-	0.2750
15	-	+	-	+	-	+	-	+	-	+	-	+	-	+	-	0.3425
16	+	+	+	+	+	+	+	+	+	+	+	+	+	+	+	0.5825

Perhaps you can see the pattern in the design. We can fit and examine a main-effects-only model:
```
lmod = smf.ols('y ~ h+d+l+b+j+f+n+a+i+e+m+c+k+g+o', speedo).fit()
lmod.sumary()
```

	coefs	stderr	tvalues	pvalues
Intercept	0.582	inf	0.00	nan
h[T.-]	-0.062	inf	-0.00	nan
d[T.-]	-0.061	inf	-0.00	nan
l[T.-]	-0.027	inf	-0.00	nan
b[T.-]	0.056	inf	0.00	nan
j[T.-]	0.001	inf	0.00	nan
f[T.-]	-0.074	inf	-0.00	nan
n[T.-]	-0.007	inf	-0.00	nan
a[T.-]	-0.068	inf	-0.00	nan
i[T.-]	-0.043	inf	-0.00	nan
e[T.-]	-0.245	inf	-0.00	nan
m[T.-]	-0.028	inf	-0.00	nan
c[T.-]	-0.090	inf	-0.00	nan
k[T.-]	-0.068	inf	-0.00	nan
g[T.-]	0.140	inf	0.00	nan
o[T.-]	-0.006	inf	-0.00	nan

n=16 p=16 Residual SD=inf R-squared=1.00

There are no degrees of freedom, because there are as many parameters as cases. We cannot do any of the usual tests. It is important to understand the coding here, so look at the X-matrix:

```
import patsy
dm = patsy.dmatrix('~ h+d+l+b+j+f+n+a+i+e+m+c+k+g+o', speedo)
np.asarray(dm)
array([[1., 1., 1., 0., 1., 0., 0., 1., 1., 0., 0., 1., 0., 1., 1., 0.],
       [1., 0., 1., 1., 1., 1., 0., 0., 1., 1., 0., 0., 0., 0., 1., 1.],
...edited...
       [1., 0., 0., 0., 0., 0., 0., 0., 0., 0., 0., 0., 0., 0., 0., 0.]])
```

We see that "+" is coded as zero and "−" is coded as one. This unnatural ordering is because of the order of "+" and "−" in the ASCII alphabet.

We do not have any degrees of freedom, so we cannot make the usual F-tests. We need a different method to determine significance. Suppose there were no significant effects and the errors were normally distributed. The estimated effects would then just be linear combinations of the errors and hence normal. We now make a normal quantile plot of the main effects with the idea that outliers represent significant effects:

```
ii = np.argsort(lmod.params[1:])
scoef = np.array(lmod.params[1:])[ii]
lcoef = (speedo.columns[:-1])[ii]
n = len(scoef)
qq = stats.norm.ppf(np.arange(1,n+1)/(n+1))
fig, ax = plt.subplots()
ax.scatter(qq, scoef,s=1)
for i in range(len(qq)):
    ax.annotate(lcoef[i], (qq[i],scoef[i]))
plt.xlabel("Normal Quantiles")
plt.ylabel("Sorted coefficients")
```

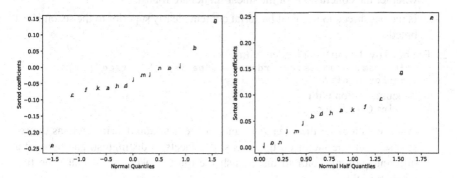

Figure 16.7   Q–Q plots of effects for speedometer cable analysis.

See Figure 16.7. Notice that "e" and possibly "g" are extreme. Since the "e" effect is negative, the "+" level of "e" increases the response. Since shrinkage is a bad thing, increasing the response is not good, so we would prefer whatever "wire braid" type corresponds to the "−" level of "e". The same reasoning for "g" leads us to expect that a larger (assuming that is "+") would decrease shrinkage.

A half-normal plot, as described in Section 6.2.1, is better for detecting extreme points:

```
n = len(scoef)
qq = stats.norm.ppf((n + np.arange(1,n+1))/(2*n+1))
acoef = np.abs(lmod.params[1:])
ii = np.argsort(acoef)
acoef = acoef[ii]
lcoef = (speedo.columns[:-1])[ii]
fig, ax = plt.subplots()
ax.scatter(qq, acoef,s=1)
for i in range(len(qq)):
    ax.annotate(lcoef[i], (qq[i],acoef[i]))
plt.xlabel("Normal Half Quantiles")
plt.ylabel("Sorted absolute coefficients")
```

We might now conduct another experiment focusing on the effects of "e" and "g."

Exercises

1. Data on the butterfat content of milk from Canadian cows of five different breeds and two different ages can be found in the butterfat dataset.

   (a) Plot the data twice, once with the breed on the x-axis and once with age on the x-axis. Which breed produces the most butterfat? Is there visual evidence of an interaction?

   (b) Determine whether there is an interaction between breed and age.

   (c) Determine whether there is a statistically significant difference between breeds and also ages.

   (d) Make the residual-fitted plot and QQ-plot for your chosen model and comment whether the conclusions of the linear model are robust.

   (e) Is the best breed in terms of butterfat content clearly superior to the second best breed?

2. The barley data may be loaded with
   ```
   df = sm.datasets.get_rdataset("barley","lattice")
   barley = df.data
   ```
   and documentation with
   ```
   print(df.__doc__)
   ```

   (a) Plot the data using the catplot function from seaborn with yield as the response. Put one factor on the x-axis, use facets to distinguish another and a different plotting color to distinguish the third. Take care to make the tick labels readable.

   (b) Were the yields in 1931 generally better? Are there any clear exceptions to this?

   (c) Fit a model with all interactions including the three way interaction. Can you test the three way interaction? (Note: You will find it necessary to rename the yield response to, say Yield to avoid an error.)

3. In this question, we will determine the important factors in the sono dataset where the Intensity is the response and the other variables are predictors.

(a) Make plots of the response against each of the predictors. Under what conditions do the highest intensities occur?

(b) If all predictors are considered, how many possible two way interactions exist? Will it be possible to estimate all of these?

(c) Fit a model with all the predictors but main effects only. Which effects are significant? What effects are large and what are small?

(d) Fit a model with only predictors with large effects from the previous model. Use an F-test to compare these two models.

(e) Fit a model with just the large-effect predictors and their two-way interactions. You should receive a warning about a singularity. How many parameters can be estimated in this model?

(f) Print out the data matrix for just the large-effect predictors. Do all possible combinations of these predictors occur?

(g) What settings should be used for the predictors that matter if maximizing intensity is the goal?

4. In the `rats` data, we model the survival time in terms of the poison and treatment.

(a) Plot the data in two different ways, each time showing both predictors. Is there evidence of interaction?

(b) Fit a model with both poison and treatment as predictors. Construct the residual-fitted plot and comment.

(c) Find a common transformation on the response that solves the problem revealed in the previous question.

(d) Test the significance of the predictors.

(e) Apply the Tukey non-additivity test for independence.

(f) What treatment and poison combination minimizes the survival time? How confident are you in this choice?

5. The `peanut` data come from a fractional factorial experiment to investigate factors that affect an industrial process using carbon dioxide to extract oil from peanuts.

(a) Make plots of the data, one predictor at a time. What tends to maximize solubility?

(b) Fit a model with main effects only. Which factors significantly increase solubility?

(c) Fit a model with all two-way interactions. What can be said about which interactions may have an effect on solubility?

(d) Fit the best model using only two predictors. What can be said about the best predictor settings to maximize solubility?

(e) The client wishes to find a more precise setting of the predictors beyond high-/low to optimize the response. What should the client do to achieve this?

6. The "High School and Beyond" data is found in `hsb`.

(a) Plot the data using boxplots for each variable with math as the response.

(b) Fit a model with the main effects only. What factors are statistically significant relative to this model?

(c) Fit a model with all the two way interactions. Which interactions are statistically significant?

(d) Which of the two models should be preferred?

(e) Make the residual-fitted plot for the main effects model and comment.

Chapter 17

Experiments with Blocks

In a completely randomized design (CRD), the treatments are assigned to the experimental units at random. This is appropriate when the units are homogeneous, as has been assumed in the designs leading to the one- and two-way analyses of variances (ANOVAs). Sometimes, we may suspect that the units are heterogeneous, but we cannot describe the form the difference takes — for example, we may know that a group of patients are not identical, but we may have no further information about them. In this case, it is still appropriate to use a CRD. Of course, the randomization will tend to spread the heterogeneity around to avoid systematic bias, but the real justification lies in the randomization test discussed in Section 3.3. Under the null hypothesis, there is no link between a factor and the response. In other words, the responses have been assigned to the units in a way that is unlinked to the factor. This corresponds to the randomization used in assigning the levels of the factor to the units. This is why the randomization is crucial because it allows us to make this argument. Now if the difference in the response between levels of the factor seems too unlikely to have occurred by chance, we can reject the null hypothesis. The normal-based inference is approximately equivalent to the permutation-based test. Since the normal-based inference is much quicker, we usually prefer to use that.

When the experimental units are heterogeneous in a known way and can be arranged into *blocks* where the within-block variation is small, but the between-block variation is large, a *block design* can be more efficient than a CRD. We prefer to have a block size equal to the number of treatments. If this cannot be done, an *incomplete* block design must be used.

Sometimes the blocks are determined by the experimenter. For example, suppose we want to compare four treatments and have 20 patients available. We might divide the patients into five blocks of four patients each where the patients in each block have some relevant similarity. We might decide this subjectively in the absence of specific information. In other cases, the blocks are predetermined by the nature of the experiment. For example, suppose we want to test three crop varieties on five fields. Restrictions on planting, harvesting and irrigation equipment might allow us only to divide the fields into three strips. The fields are the blocks.

In a randomized block design, the treatment levels are assigned randomly within a block. This means the randomization is restricted relative to the full randomization used in the CRD. This has consequences for the inference. There are fewer possible permutations for the random assignment of the treatments; therefore, the computation of the significance of a statistic based on the permutation test would need to

273

be modified. Consequently, a block effect must be included in the model used for inference about the treatments, even if the block effect turns out not to be significant.

17.1 Randomized Block Design

We have one treatment factor, τ at t levels and one blocking factor, β at r levels. The model is:

$$y_{ij} = \mu + \tau_i + \beta_j + \varepsilon_{ij}$$

where τ_i is the treatment effect and ρ_j is the blocking effect. There is one observation on each treatment in each block. This is called a randomized complete block design (RCBD). The analysis is then very similar to a two-factor experiment with no replication. We have a limited ability to detect an interaction between treatment and block. We can check for a treatment effect. We can also check the block effect, but this is only useful for future reference. Blocking is a feature of the experimental units and restricts the randomized assignment of the treatments. This means that we cannot regain the degrees of freedom devoted to blocking even if the blocking effect is not significant.

Load the packages:

```
import pandas as pd
import numpy as np
import matplotlib.pyplot as plt
import statsmodels.api as sm
import statsmodels.formula.api as smf
import seaborn as sns
from scipy import stats
import faraway.utils
```

We illustrate this with an experiment to compare eight varieties of oats. The growing area was heterogeneous and so was grouped into five blocks. Each variety was sown once within each block and the yield in grams per 16-ft row was recorded. The data come from Anderson and Bancroft (1952).

We start with a look at the data:

```
import faraway.datasets.oatvar
oatvar = faraway.datasets.oatvar.load()
oatvar.pivot(index = 'variety', columns='block', values='yield')
```

block	I	II	III	IV	V
variety					
1	296	357	340	331	348
2	402	390	431	340	320
3	437	334	426	320	296
4	303	319	310	260	242
5	469	405	442	487	394
6	345	342	358	300	308
7	324	339	357	352	220
8	488	374	401	338	320

Color plots of the data can be obtained in seaborn by:

```
sns.boxplot(x="variety", y="yield", data=oatvar)
sns.boxplot(x="block", y="yield", data=oatvar)
sns.catplot(x='variety',y='yield',hue='block', data=oatvar,
    kind='point')
sns.catplot(x='block',y='yield',hue='variety', data=oatvar,
```

```
    kind='point')
```
Alternatively, we can use `matplotlib`
```
faclevels = np.unique(oatvar.block)
oatvar["variety"] = oatvar["variety"].astype('category')
oatvar["block"] = oatvar["block"].astype('category')
for i in np.arange(len(faclevels)):
    j = (oatvar.block == faclevels[i])
    plt.plot(oatvar.variety.cat.codes[j], oatvar.grams[j])
    plt.annotate(faclevels[i],
        [-0.2,np.ravel(oatvar.grams[j])[0]])
    plt.annotate(faclevels[i],
        [7.05,np.ravel(oatvar.grams[j])[7]])
plt.xlabel("Variety")
plt.xticks(np.arange(0,8),np.arange(1,9))
plt.ylabel("Yield")
faclevels = np.unique(oatvar.variety)
for i in np.arange(len(faclevels)):
    j = (oatvar.variety == faclevels[i])
    plt.plot(oatvar.block.cat.codes[j], oatvar.grams[j])
    plt.annotate(faclevels[i],
                [-0.15,np.ravel(oatvar.grams[j])[0]])
    plt.annotate(faclevels[i],
                [4.05,np.ravel(oatvar.grams[j])[4]])
plt.xlabel("Block")
plt.xticks(np.arange(0,5),oatvar.block.cat.categories)
plt.ylabel("Yield")
```
See Figure 17.1. There is no indication of outliers, skewness or non-constant vari-

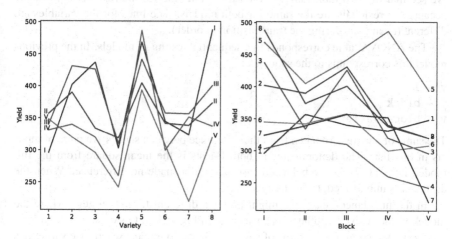

Figure 17.1   Interaction plots of oat variety data.

ance. Nor is there any clear evidence of an interaction effect. We can move onto the modeling:
```
lmod = smf.ols('grams ~ variety + block', oatvar).fit()
sm.stats.anova_lm(lmod).round(3)
            df      sum_sq    mean_sq      F   PR(>F)
variety    7.0   77523.575  11074.796  8.284   0.000
```

```
block      4.0  33395.500  8348.875  6.245   0.001
Residual  28.0  37433.300  1336.904    NaN     NaN
```

We have used grams as the response which is identical to yield. Unfortunately, yield is a keyword in Python which has a particular meaning. You will receive a perplexing error if you use it as a variable in some statistical models. Be warned.

We see that both effects are significant. It would not matter if we transposed the order of variety and block in the model. The balanced design ensures that we get the same outcome. Try this and verify that this does not change the ANOVA table (apart from transposing the lines for the variables).

Now suppose we unfortunately lost the first observation in the dataset:

```
oat1 = oatvar.iloc[1:,]
lmod = smf.ols('grams ~ variety + block', oat1).fit()
sm.stats.anova_lm(lmod).round(4)
            df      sum_sq     mean_sq       F    PR(>F)
variety    7.0  75901.6308  10843.0901  9.4538   0.0000
block      4.0  38017.9080   9504.4770  8.2867   0.0002
Residual  27.0  30967.6920   1146.9516     NaN      NaN
```

Now try reversing the order of the variables in the model:

```
lmod = smf.ols('grams ~ block + variety', oat1).fit()
sm.stats.anova_lm(lmod).round(4)
            df      sum_sq     mean_sq       F    PR(>F)
block      4.0  38580.6415   9645.1604  8.4094   0.0002
variety    7.0  75338.8973  10762.6996  9.3837   0.0000
Residual  27.0  30967.6920   1146.9516     NaN      NaN
```

We see that the two tables are not the same. The line for the residual is the same because the residuals are the same for both models. The lines for the variables are different (even considering the transposition in order).

The ANOVA table corresponds to a sequential testing of models. In the previous model this corresponds to the sequence:

y ~ 1
y ~ block
y ~ block+variety

The sums of squares are taken from the difference between successive pairs of models in this list. The denominator in both F-tests is the mean square from the full model, here 1147. In the balanced case, the order made no difference. When the designed is unbalanced, the order matters.

In the unbalanced case, we might prefer a different testing strategy where the model of alternative hypothesis is always the full model, in this case, y ~ block + variety. We compute the sums of squares relative to models where one variable is omitted from this full model. We can obtain the ANOVA table corresponding to this strategy with:

```
sm.stats.anova_lm(lmod,typ=3).round(4)
                sum_sq    df         F    PR(>F)
Intercept  358745.3601   1.0  312.7816   0.0000
block       38017.9080   4.0    8.2867   0.0002
variety     75338.8973   7.0    9.3837   0.0000
Residual    30967.6920  27.0       NaN      NaN
```

In SAS, this is known as "Type 3 sum of squares" while the default (sequential) table is known as "Type 1 sum of squares". We would prefer a more descriptive terminology — "drop one" vs. "sequential" for example.

In this example, we would prefer the drop one strategy. When we test for a variety effect, we would want to first allow for a possible block effect (and vice versa). Here, the variables are significant enough that either choice produces the same answer but this will not always be true. Remember that if you have a balanced design, the distinction will not matter, as both methods will produce the same outcome.

Check the diagnostics, as seen in Figure 17.2:

```
lmod = smf.ols('grams ~ variety + block', oatvar).fit()
plt.scatter(lmod.fittedvalues, lmod.resid)
plt.axhline(0)
plt.xlabel("Fitted values")
plt.ylabel("Residuals")
p=sm.qqplot(lmod.resid, line="q")
```

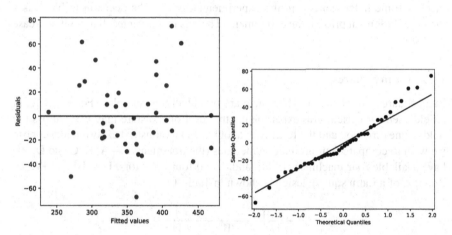

Figure 17.2   Diagnostic plots for the oat variety data.

An examination of which varieties give the highest yields and which are significantly better than others can now follow.

We did assume that the interactions were not significant. We looked at the interaction plots, but we can also execute the Tukey nonadditivity test:

```
varcoefs = np.append([0.],lmod.params[1:8])
varcoefs = np.repeat(varcoefs,5)
blockcoefs = np.append([0.],lmod.params[8:12])
blockcoefs = np.tile(blockcoefs,8)
oatvar['crossp'] = varcoefs * blockcoefs
tmod = smf.ols("grams ~ variety + block + crossp", oatvar).fit()
sm.stats.anova_lm(tmod).round(3)
```

	df	sum_sq	mean_sq	F	PR(>F)
variety	7.0	77523.575	11074.796	8.172	0.000
block	4.0	33395.500	8348.875	6.161	0.001
crossp	1.0	843.960	843.960	0.623	0.437
Residual	27.0	36589.340	1355.161	NaN	NaN

Because the p-value of the treatment times block effect is 0.437, we accept the null hypothesis of no interaction. Of course, the interaction may be of a nonmultiplicative form, but there is little we can do about that.

Relative advantage of RCBD over CRD

We can measure precision by considering var $\hat{\tau}$ or equivalently $\hat{\sigma}^2$. We should compare the $\hat{\sigma}^2$ for designs with the same sample size. We define *relative efficiency* as $\hat{\sigma}^2_{CRD}/\hat{\sigma}^2_{RCBD}$ where the quantities can be computed by fitting models with and without the blocking effect. For the example above:

```
lmcrd = smf.ols("grams ~ variety", oatvar).fit()
lmcrd.scale/lmod.scale
1.6556
```

So a CRD would require 66% more observations to obtain the same level of precision as an RCBD.

The efficiency is not guaranteed to be greater than one. Only use blocking where there is some heterogeneity in the experimental units. The decision to block is a matter of judgment prior to the experiment. There is no guarantee that it will increase precision.

17.2 Latin Squares

Latin squares can be useful when there are two blocking variables. For example, in a field used for agricultural experiments, the level of moisture may vary across the field in one direction and the fertility in another. In an industrial experiment, suppose we wish to compare four production methods (the treatments) — A, B, C and D. We have available four machines, 1, 2, 3 and 4, and four operators, I, II, III and IV. An example of a Latin square design is shown in Table 17.1.

	1	2	3	4
I	A	B	C	D
II	B	D	A	C
III	C	A	D	B
IV	D	C	B	A

Table 17.1   A 4×4 Latin square showing the treatments (A to D) used for different combinations of two factors.

Each treatment is assigned to each block once and only once. We should choose randomly from all the possible Latin square layouts.

Let τ be the treatment factor and β and γ be the two blocking factors; then the model is:

$$y_{ijk} = \mu + \tau_i + \beta_j + \gamma_k + \varepsilon_{ijk} \qquad i, j, k = 1, \ldots, t$$

All combinations of i, j and k do not appear. To test for a treatment effect, fit a model without the treatment effect and compare using the F-test. The Tukey pairwise CIs are:

$$\hat{\tau}_l - \hat{\tau}_m \pm \frac{q_{t,(t-1)(t-2)}}{\sqrt{2}} \hat{\sigma} \sqrt{2/t}$$

The Latin square can be even more efficient than the RCBD provided that the blocking effects are sizable. There are some variations on the Latin square. The Latin square can be replicated if more runs are available. We need to have both block sizes to be equal to the number of treatments. This may be difficult to achieve. Latin rectangle designs are possible by adjoining Latin squares. When there are three blocking variables, a Graeco–Latin square may be used but these rarely arise in practice.

The Latin square can also be used for comparing three treatment factors. Only t^2 runs are required compared to the t^3 required if all combinations were run. (The downside is that you cannot estimate the interactions if they exist.) This is an example of a *fractional factorial* as discussed in Section 16.4.

In an experiment reported by Davies (1954), four materials, A, B, C and D, were fed into a wear-testing machine. The response is the loss of weight in 0.1 mm over the testing period. The machine could process four samples at a time, and past experience indicated that there were some differences due to the positions of these four samples. Also some differences were suspected from run to run. Four runs were made. We can display the layout of the data:

```
import faraway.datasets.abrasion
abrasion = faraway.datasets.abrasion.load()
abrasion.pivot(index = 'run', columns='position', values='wear')
```

position	1	2	3	4
run				
1	235	236	218	268
2	251	241	227	229
3	234	273	274	226
4	195	270	230	225

We can view the Latin square structure corresponding to this table:

```
abrasion.pivot(index = 'run', columns='position',
               values='material')
```

position	1	2	3	4
run				
1	C	D	B	A
2	A	B	D	C
3	D	C	A	B
4	B	A	C	D

Plot the data in color with seaborn (not shown)

```
sns.catplot(x='run',y='wear',hue='position', data=abrasion,
    kind='point',scale=0.5)
sns.catplot(x='run',y='wear',hue='material', data=abrasion,
    kind='point',scale=0.5)
```

Alternatively with matplotlib:

```
faclevels = np.unique(abrasion.position)
lineseq = ['-','--','-.',':']
for i in np.arange(len(faclevels)):
    j = (abrasion.position == faclevels[i])
    plt.plot(abrasion.run[j], abrasion.wear[j],
             lineseq[i],label=faclevels[i])
plt.legend(title="Position")
plt.xlabel("Run")
plt.xticks(np.arange(1,5))
plt.ylabel("Wear")
```

and
```
faclevels = np.unique(abrasion.run)
lineseq = ['-','--','-.',':']
for i in np.arange(len(faclevels)):
    j = (abrasion.run == faclevels[i])
    plt.plot(abrasion.position[j], abrasion.wear[j],
             lineseq[i],label=faclevels[i])
plt.legend(title="Run")
plt.xlabel("Position")
plt.xticks(np.arange(1,5))
plt.ylabel("Wear")
```

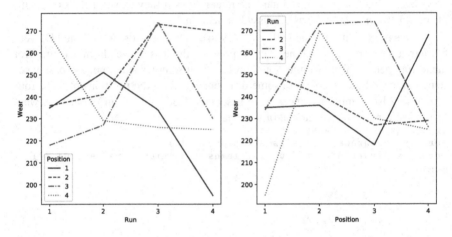

Figure 17.3   Amount of wear depending on material, run and position.

Examine the plots in Figure 17.3. There appear to be differences in all variables. No outliers, skewness or unequal variance are apparent. There is no obvious sign of interaction.

Now fit the Latin square model and test each variable relative to the full model. We need to declare run and position as categorical variables so that they are not treated as numbers.
```
abrasion["run"] = abrasion["run"].astype('category')
abrasion["position"] = abrasion["position"].astype('category')
lmod = smf.ols('wear ~ run + position + material', abrasion).fit()
sm.stats.anova_lm(lmod,typ=3).round(3)
```

	df	sum_sq	mean_sq	F	PR(>F)
run	3.0	986.5	328.833	5.369	0.039
position	3.0	1468.5	489.500	7.992	0.016
material	3.0	4621.5	1540.500	25.151	0.001
Residual	6.0	367.5	61.250	NaN	NaN

We see that all variables are statistically significant. There are clear differences between the materials. We checked the diagnostics on the model, which showed nothing remarkable. We examine the coefficients:
```
lmod.sumary()
```

	coefs	stderr	tvalues	pvalues
Intercept	254.750	6.187	41.17	0.0000

```
run[T.2]        -2.250  5.534   -0.41  0.6984
run[T.3]        12.500  5.534    2.26  0.0647
run[T.4]        -9.250  5.534   -1.67  0.1457
position[T.2]   26.250  5.534    4.74  0.0032
position[T.3]    8.500  5.534    1.54  0.1755
position[T.4]    8.250  5.534    1.49  0.1866
material[T.B]  -45.750  5.534   -8.27  0.0002
material[T.C]  -24.000  5.534   -4.34  0.0049
material[T.D]  -35.250  5.534   -6.37  0.0007
```

n=16 p=10 Residual SD=7.826 R-squared=0.95

We see that material B looks best (in terms of least wear) followed by material D. Is the difference significant though? Which materials in general are significantly better than others? We need the Tukey pairwise intervals to help determine this. The width of the band is calculated in the usual manner:

```
from statsmodels.sandbox.stats.multicomp import get_tukeyQcrit
get_tukeyQcrit(4,6)*lmod.bse[1]/np.sqrt(2)
19.174
```

The width of the interval is 19.1. We can make a table of the material differences:

```
treatname = 'material'
treatlevs = ['A','B','C','D']
```

and

```
ii = lmod.params.index.str.match(treatname)
mcoefs = np.append([0],lmod.params[ii])
import itertools
dp = set(itertools.combinations(range(0,len(treatlevs)),2))
dcoef = []
namdiff = []
for cp in dp:
    dcoef.append(mcoefs[cp[0]] - mcoefs[cp[1]])
    namdiff.append(treatlevs[cp[0]] + '-' + treatlevs[cp[1]])
thsd = pd.DataFrame({'Difference':dcoef},index=namdiff)
cvband = get_tukeyQcrit(len(treatlevs),
         lmod.df_resid) * lmod.bse[1]/np.sqrt(2)
thsd["lb"] = thsd.Difference - cvband
thsd["ub"] = thsd.Difference + cvband
thsd.round(2)
     Difference      lb     ub
A-B       45.75   26.58  64.92
B-C      -21.75  -40.92  -2.58
B-D      -10.50  -29.67   8.67
C-D       11.25   -7.92  30.42
A-D       35.25   16.08  54.42
A-C       24.00    4.83  43.17
```

We see that the (B, D) and (D, C) differences are not significant at the 5% level, but that all the other differences are significant.

If maximizing resistance to wear is our aim, we would pick material B but if material D offered a better price, we might have some cause to consider switching to D. The decision would need to be made with cost–quality trade-offs in mind.

Now we compute how efficient the Latin square is compared to other designs. We compare to the completely randomized design:

```
lmodr = smf.ols('wear ~ material', abrasion).fit()
lmodr.scale/lmod.scale
```

3.8401

We see that the Latin square is 3.84 times more efficient than the CRD. This is a substantial gain in efficiency. The Latin square may also be compared to designs where we block only one of the variables. The efficiency relative to these designs is less impressive, but still worthwhile.

17.3 Balanced Incomplete Block Design

In a complete block design, the block size is equal to the number of treatments. When the block size is less than the number of treatments, an incomplete block design must be used. For example, in the oat data, suppose six oat varieties were to be compared, but each field had space for only four plots.

In an incomplete block design, the treatments and blocks are *not* orthogonal. Some treatment contrasts will not be identifiable from certain block contrasts. This is an example of *confounding*. This means that those treatment contrasts effectively cannot be examined. In a balanced incomplete block (BIB) design, all the pairwise differences are identifiable and have the same standard error. Pairwise differences are more likely to be interesting than other contrasts, so the design is constructed to facilitate this.

Suppose, we have four treatments ($t = 4$) designated A, B, C, D and the block size, $k = 3$ and there are $b = 4$ blocks. Therefore, each treatment appears $r = 3$ times in the design. One possible BIB design is:

Block 1	A	B	C
Block 2	A	B	D
Block 3	A	C	D
Block 4	B	C	D

Table 17.2   BIB design for four treatments with four blocks of size 3.

Each pair of treatments appears in the same block $\lambda = 2$ times — this feature means simpler pairwise comparison is possible. For a BIB design, we require:

$$b \geq t > k$$
$$rt = bk = n$$
$$\lambda(t-1) = r(k-1)$$

This last relation holds because the number of pairs in a block is $k(k-1)/2$ so the total number of pairs must be $bk(k-1)/2$. On the other hand, the number of treatment pairs is $t(t-1)/2$. The ratio of these two quantities must be λ.

Since λ has to be an integer, a BIB design is not always possible even when the first two conditions are satisfied. For example, consider $r = 4, t = 3, b = 6, k = 2$ and then $\lambda = 2$ which is OK, but if $r = 4, t = 4, b = 8, k = 2$, then $\lambda = 4/3$ so no BIB is possible. (Something called a partially balanced incomplete block design can then be used.) BIBs are also useful for competitions where not all contestants can fit in the same race.

The model we fit is the same as for the RCBD:

$$y_{ij} = \mu + \tau_i + \beta_j + \varepsilon_{ij}$$

In our example, a nutritionist studied the effects of six diets, "a" through "f," on weight gains of domestic rabbits. From past experience with sizes of litters, it was felt that only three uniform rabbits could be selected from each available litter. There were ten litters available forming blocks of size 3. The data come from Lentner and Bishop (1986). Examine the data:

```
import faraway.datasets.rabbit
rabbit = faraway.datasets.rabbit.load()
pt = rabbit.pivot(index = 'treat', columns='block', values='gain')
pt.replace(np.nan," ", regex=True)
block    b1    b10    b2    b3    b4    b5    b6    b7    b8    b9
treat
a             37.3  40.1        44.9                    45.2   44
b      32.6         38.1                          37.3  40.6        30.6
c      35.2         40.9  34.6  43.9  40.9
d             42.3        37.5        37.3        37.9              27.5
e                               40.8    32  40.5        38.5  20.6
f      42.2  41.7        34.3                    42.8        51.9
```

The empty cells correspond to no observation. The BIB structure is apparent — each pair of diets appears in the same block exactly twice. Now plot the data, as seen in Figure 17.4. In the first plot, we can distinguish the six levels of treatment with a different plotting character. In the second plot, it is difficult to distinguish the ten levels of block with different shapes, so we have used a simple dot. Some jittering is necessary to distinguish the points. There is nothing remarkable about the content of the plots as we see no evidence of outliers, skewness or nonconstant variance.

Color plots of the data are available in seaborn using (not shown):

```
sns.swarmplot(x='block',y='gain',hue='treat',data=rabbit)
plt.legend(bbox_to_anchor=(1.05, 1), loc=2, borderaxespad=0.)
```

and

```
sns.swarmplot(x='treat',y='gain',hue='block',data=rabbit)
plt.legend(bbox_to_anchor=(1.05, 1), loc=2, borderaxespad=0.)
```

We plot the data using matplotlib with:

```
rabbit["treat"] = rabbit["treat"].astype('category')
rabbit["block"] = rabbit["block"].astype('category')
plt.scatter(rabbit.block.cat.codes,rabbit.gain,color="white")
for i in np.arange(0,len(rabbit.gain)):
    plt.annotate(rabbit.treat.iloc[i],
        (rabbit.block.cat.codes.iloc[i],rabbit.gain.iloc[i]))
plt.xticks(np.arange(0,10),rabbit.block.cat.categories)
plt.ylabel("Gain")
plt.xlabel("Block")
plt.scatter(rabbit.treat.cat.codes,rabbit.gain,color="white")
for i in np.arange(0,len(rabbit.gain)):
    plt.annotate(rabbit.block.iloc[i],
        (rabbit.treat.cat.codes.iloc[i],rabbit.gain.iloc[i]))
plt.xticks(np.arange(0,6),rabbit.treat.cat.categories)
plt.ylabel("Gain")
plt.xlabel("Treatment")
```

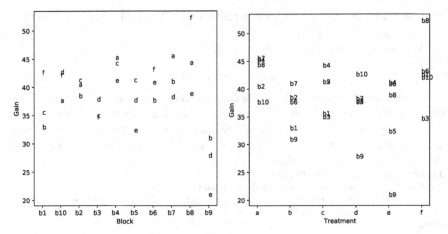

Figure 17.4   Strip plots of rabbit diet data.

We fit the model and test the significance of the effects:
```
lmod = smf.ols('gain ~ treat + block', rabbit).fit()
sm.stats.anova_lm(lmod,typ=3).round(3)
```

	sum_sq	df	F	PR(>F)
Intercept	1945.500	1.0	193.553	0.000
treat	158.727	5.0	3.158	0.038
block	595.735	9.0	6.585	0.001
Residual	150.773	15.0	NaN	NaN

We use the drop one (Type 3) option here since to test either term, we need the other term to be present in the model.

Now check the diagnostics:
```
sns.residplot(lmod.fittedvalues, lmod.resid)
plt.xlabel("Fitted values")
plt.ylabel("Residuals")
p=sm.qqplot(lmod.resid, line="q")
```
The plots are not shown, as they show nothing of interest.

Now we check which treatments differ. The Tukey pairwise CIs need to be directly constructed because this is not a complete layout. We extract the information about the treatment effects and the standard error:
```
lmod.sumary()
```

	coefs	stderr	tvalues	pvalues
Intercept	36.014	2.589	13.91	0.0000
treat[T.b]	-1.742	2.242	-0.78	0.4493
treat[T.c]	0.400	2.242	0.18	0.8608
treat[T.d]	0.067	2.242	0.03	0.9767
treat[T.e]	-5.225	2.242	-2.33	0.0341
treat[T.f]	3.300	2.242	1.47	0.1617
block[T.b10]	3.297	2.796	1.18	0.2567
block[T.b2]	4.133	2.694	1.53	0.1458
block[T.b3]	-1.803	2.694	-0.67	0.5136
block[T.b4]	8.794	2.796	3.15	0.0067
block[T.b5]	2.306	2.796	0.82	0.4225

```
block[T.b6]    5.408   2.694    2.01  0.0631
block[T.b7]    5.778   2.796    2.07  0.0565
block[T.b8]    9.428   2.796    3.37  0.0042
block[T.b9]   -7.481   2.796   -2.68  0.0173
```

```
n=30 p=15 Residual SD=3.170 R-squared=0.86
```

We see that the standard error for the pairwise comparison of treatments is 2.24. Notice that all the treatment standard errors are equal because of the BIB. Now compute the Tukey confidence intervals

```
mcoefs = np.append([0],lmod.params[1:6])
nmats = [chr(i) for i in range(ord('a'),ord('f')+1)]
p = len(mcoefs)
dp = set(itertools.combinations(range(0,p),2))
dcoef = []
namdiff = []
for cp in dp:
    dcoef.append(mcoefs[cp[0]] - mcoefs[cp[1]])
    namdiff.append(nmats[cp[0]] + '-' + nmats[cp[1]])
thsd = pd.DataFrame({'Difference':dcoef},index=namdiff)
thsd["lb"] = thsd.Difference - \
    get_tukeyQcrit(p,lmod.df_resid) * lmod.bse[1]/np.sqrt(2)
thsd["ub"] = thsd.Difference + \
    get_tukeyQcrit(p,lmod.df_resid) * lmod.bse[1]/np.sqrt(2)
thsd.round(2)
```

	Difference	lb	ub
a-b	1.74	-5.53	9.02
b-c	-2.14	-9.42	5.13
b-d	-1.81	-9.08	5.47
e-f	-8.53	-15.80	-1.25
b-e	3.48	-3.79	10.76
b-f	-5.04	-12.32	2.23
c-e	5.62	-1.65	12.90
a-f	-3.30	-10.58	3.98
c-d	0.33	-6.94	7.61
c-f	-2.90	-10.18	4.38
a-e	5.23	-2.05	12.50
a-d	-0.07	-7.34	7.21
d-e	5.29	-1.98	12.57
a-c	-0.40	-7.68	6.88
d-f	-3.23	-10.51	4.04

Only the $e-f$ difference is significant.

Now let's see how much better this blocked design is than the CRD. We compute the relative efficiency:

```
lmodr = smf.ols('gain ~ treat', rabbit).fit()
lmodr.scale/lmod.scale
3.0945
```

Blocking was well worthwhile here.

Exercises

1. The alfalfa data arise from a Latin square design where the treatment factor is inoculum and the blocking factors are shade and irrigation.

   (a) Display the data as a table that makes the Latin square structure apparent.

(b) Make three plots of the data using the same response but varying the predictor. Comment on whether the assumptions of linear modeling appear to be justified.

(c) Fit the appropriate linear model and display the ANOVA table. Which factors are significant? Does it matter which type of ANOVA table is produced?

(d) Compute the Tukey pairwise comparison intervals for the inoculum. Is E clearly the worst choice?

(e) Compute the relative efficiency compared to the CRD. Compute the relative efficiency compared to a design that blocked only on shade. Should we block on irrigation in future experiments?

2. The `eggprod` comes from a randomized block experiment to determine factors affecting egg production.

(a) Plot the data with eggs as the response. Is there evidence of an interaction effect?

(b) Fit the RBD model and use a residual-fitted plot to check the model assumptions.

(c) Is there a significant difference between the treatments? Is there a significant difference between the blocks?

(d) Use the Tukey method to test for an interaction.

(e) How many pairwise comparisons exist among the treatments? Use Bonferroni correction to determine whether any of the treatments may be distinguished.

(f) Compute the relative efficiency of this design compared to a CRD. Would it be worth using the same blocking design in future experiments?

3. The `morley` data comes from a 19th century experiment to measure the speed of light. Use the `sm.datasets.get_rdataset` to obtain the data and description.

(a) Plot the data with Speed as the response with a view to check whether there is any trend within runs over the different experiments.

(b) Fit a linear model with speed as the response and run and experiment both in the numeric forms. Examine the ANOVA output and the regression summary output. Are the p-values the same? Explain.

(c) Convert the experiment variable to a factor. Why might this be reasonable? Perform the appropriate test to determine whether there is a significant difference between experiments.

(d) Now change the run to a factor and test for the significance of both factors.

(e) As it happens, Morley collected all 100 observations in the order seen but were later divided into 'experiments'. Plot the data in run order and comment on whether there is a trend.

4. The `OrchardSprays` data come from an experiment that was conducted to assess the potency of various constituents of orchard sprays in repelling honeybees. Use the `sm.datasets.get_rdataset` to obtain the data and description.

(a) Display the data in a table that reveals the Latin square design. Suppose we permuted the levels of one of the factors. Would this remain a Latin square design?

(b) The treatment levels are in a particular order. Plot the data to reveal whether there is in any trend with this order that might be related to the other two factors.

(c) Fit a model with decrease as the response and treatment, row and column position as predictors. Keep the row and column variables numeric. What predictors are significant?

(d) Create factor versions of the row and column variables. Refit the model and determine which factors are significant. Comment on any differences with the previous result.

(e) Use a test to compare the models of the previous two questions. What should we conclude?

(f) Construct a model where the treatment is considered ordinal. Compare it to a model where it is considered nominal. Hint: You may find the ord function useful.

(g) Given the objective of trying to minimize the decrease, what recommendation should be made for the level of sulphur?

5. The resceram data arise from an experiment to test the effects of resistor shape on current noise. The resistors are mounted on plates and only three resistors will fit on a plate although there are four different shapes to be tested.

(a) Display the data in a table that reveals the structure of the design. How many times does each pair of shapes occur within the same plate?

(b) Plot the data and comment.

(c) Fit a model with both predictors. Compute the ANOVA table using both sequential (Type 1) and drop one (Type 3) sums of squares. Which one should you use and what is the conclusion?

(d) All things being equal, which factor levels would you recommend to minimize the noise?

(e) Compute the simultaneous 95% pairwise comparisons for the shapes. Can we conclude that one shape is clearly the best for minimising noise?

6. The penicillin data derive from the production of penicillin which uses a raw material, corn steep liquor, that is quite variable and can only be made in blends sufficient for four runs. There are four processes, A, B, C and D, for the production.

(a) Make two plots of the data where the treatment and blend are plotted differently. What can you conclude?

(b) Fit an RBD model and test for the significance of the factors. Remember to rename the response.

(c) Make a residual-fitted plot and a QQ-plot. What do you conclude?

(d) Suppose another analyst decides to ignore the blend information. Compare estimates and standard errors of the treatment effects in the two models. Which is the best approach?

(e) Compute the relative efficiency of this design over the CRD. What sample size would a CRD need to be in order to achieve the same precision?

7. Is the standard Sudoku solution a Latin square? Does the Sudoku solution contain any additional features that might be useful for design purposes? Propose a situation where it could be helpful.

Appendix A

About Python

Installation

There are several ways to install Python. I used the distribution from Anaconda at www.anaconda.com. I also used Jupyter notebooks from jupyter.org for my working environment. Although I can recommend these choices, there are many reasonable alternatives which should work equally well with the code in this book. I have used Python 3 exclusively.

How to Learn Python

The Internet provides extensive resources for learning Python. You will want to find resources that focus on the manipulation and plotting of data rather than programming. Books and guides that promise to teach you to do data science with Python will be most helpful. I have assumed a basic knowledge of Python for readers of this text, but it is not necessary to learn it in detail before starting on this book. Many readers are able to pick up the language from the examples in the text with reference to the help pages. For example, for more information on the `smf.ols` function, type within Python:

```
?smf.ols
```

More advanced users may find online forums such as StackExchange helpful for more difficult questions.

Packages

This book uses some functions and data that are not part of base Python. These are collected in *packages* which you may need to install. I use:

- numpy
- scipy
- pandas
- statsmodels
- matplotlib
- seaborn
- scikit-learn
- patsy

These packages come pre-installed with Anaconda release.

You will also need to install my package `faraway` from PyPi at pypi.org typically using the command:

```
pip install faraway
```

The package provides mostly data but also a few functions.

Appearance

It is my intention that the code in the book will allow you to reproduce the output. There are a few exceptions to this. I have edited the output in some places to remove excessive output or to improve the formatting. In particular, it appears impossible to control the printing of significant digits in a global way. One rarely wants to see 3.234561445 where 3.23 would suffice. Rather than encumber the code with formatting commands, I have sometimes just rounded the output manually. Of course, we should not reduce the precision with which these numbers are internally stored.

Updates and Errata

The code and output shown in this book were generated under Python version 3.7. Python and its packages are regularly updated and improved, so more recent versions may show some differences in the output. Sometimes these changes can break the code in the text. This is more likely to occur in the additional packages. Please refer to the book website at `https://julianfaraway.github.io/LMP/` for information about such changes or errors found in the text.

Reproducible Research

It is important that you are able to reproduce your analysis in the future if needed. Sometimes, other people will need to verify your findings or, more likely, you yourself will need to modify or update your analysis. At a minimum, you should maintain a commented set of Python commands that will reproduce your analysis. I recommend Jupyter notebooks for this purpose.

Bibliography

Akaike, H. (1974). A new look at the statistical model identification. *Automatic Control, IEEE Transactions on 19*(6), 716–723.

Anderson, C. and R. Loynes (1987). *The Teaching of Practical Statistics*. New York: Wiley.

Anderson, R. and T. Bancroft (1952). *Statistical Theory in Research*. New York: McGraw-Hill.

Andrews, D. and A. Herzberg (1985). *Data: A Collection of Problems from Many Fields for the Student and Research Worker*. New York: Springer-Verlag.

Anscombe, F. J. (1973). Graphs in statistical analysis. *The American Statistician 27*(1), 17–21.

Benjamini, Y. and Y. Hochberg (1995). Controlling the false discovery rate: a practical and powerful approach to multiple testing. *Journal of the Royal Statistical Society. Series B (Methodological) 57*, 289–300.

Berkson, J. (1950). Are there two regressions? *Journal of the American Statistical Association 45*, 165–180.

Box, G., S. Bisgaard, and C. Fung (1988). An explanation and critique of Taguchi's contributions to quality engineering. *Quality and Reliability Engineering International 4*, 123–131.

Box, G., W. Hunter, and J. Hunter (1978). *Statistics for Experimenters*. New York: Wiley.

Burnham, K. P. and D. R. Anderson (2002). *Model Selection and Multi-Model Inference: A Practical Information-Theoretic Approach*. New York: Springer.

Cook, J. and L. Stefanski (1994). Simulation–extrapolation estimation in parametric measurement error models. *Journal of the American Statistical Association 89*, 1314–1328.

Davies, O. (1954). *The Design and Analysis of Industrial Experiments*. New York: Wiley.

Davison, A. and D. Hinkley (1997). *Bootstrap Methods and their Application*. Cambridge: Cambridge University Press.

de Boor, C. (2002). *A Practical Guide to Splines*. New York: Springer.

de Jong, S. (1993). SIMPLS: An alternative approach to partial least squares regression. *Chemometrics and Intelligent Laboratory Systems 18*, 251–263.

Draper, N. and H. Smith (1998). *Applied Regression Analysis* (3rd ed.). New York: Wiley.

Efron, B., T. Hastie, I. Johnstone, and R. Tibshirani (2004). Least angle regression. *Annals of Statistics 32*, 407–499.

Efron, B. and R. Tibshirani (1993). *An Introduction to the Bootstrap*. London: Chapman & Hall.

Faraway, J. (1992). On the cost of data analysis. *Journal of Computational and Graphical Statistics 1*, 215–231.

Faraway, J. (1994). Order of actions in regression analysis. In P. Cheeseman and W. Oldford (Eds.), *Selecting Models from Data: Artificial Intelligence and Statistics IV*, pp. 403–411. New York: Springer-Verlag.

Faraway, J. (2014). *Linear Models with R* (2 ed.). London: Chapman & Hall.

Faraway, J. J. (2016). Does data splitting improve prediction? *Statistics and Computing 26*(1-2), 49–60.

Fisher, R. (1936). Has Mendel's work been rediscovered? *Annals of Science 1*, 115–137.

Frank, I. and J. Friedman (1993). A statistical view of some chemometrics tools. *Technometrics 35*, 109–135.

Freedman, D. and D. Lane (1983). A nonstochastic interpretation of reported significance levels. *Journal of Business and Economic Statistics 1*(4), 292–298.

Galton, F. (1886). Regression towards mediocrity in hereditary stature. *The Journal of the Anthropological Institute of Great Britain and Ireland 15*, 246–263.

Garthwaite, P. (1994). An interpretation of partial least squares. *Journal of the American Statistical Association 89*, 122–127.

Gelman, A. (2008). Scaling regression inputs by dividing by two standard deviations. *Statistics in Medicine 27*(15), 2865–2873.

Hamada, M. and J. Wu (2000). *Experiments: Planning, Analysis, and Parameter Design Optimization*. New York: Wiley.

Harrell, F. E. (2015). *Regression modeling strategies: with applications to linear models, logistic and ordinal regression, and survival analysis* (2 ed.). New York: Springer.

Herron, M., W. M. Jr, and J. Wand (2008). Voting Technology and the 2008 New Hampshire Primary. *Wm. & Mary Bill Rts. J. 17*, 351–374.

Hill, A. B. (1965). The environment and disease: association or causation? *Proceedings of the Royal Society of Medicine 58*(5), 295.

Hsu, J. (1996). *Multiple Comparisons Procedures: Theory and Methods*. London: Chapman & Hall.

John, P. (1971). *Statistical Design and Analysis of Experiments*. New York: Macmillan.

Johnson, M. and P. Raven (1973). Species number and endemism: the Galápagos Archipelago revisited. *Science 179*, 893–895.

Johnson, R. (1996). Fitting percentage of body fat to simple body measurements. *Journal of Statistics Education 4*(1), 265–266.

Joliffe, I. (2002). *Principal Component Analysis* (2 ed.). New York: Springer Verlag.

Jones, P. and M. Mann (2004). Climate over past millennia. *Reviews of Geophysics 42*, 1–42.

Lentner, M. and T. Bishop (1986). *Experimental Design and Analysis*. Blacksburg, VA: Valley Book Company.

Little, R. and D. Rubin (2002). *Statistical Analysis with Missing Data* (2nd ed.). New York: Wiley.

Makridakis, S., E. Spiliotis, and V. Assimakopoulos (2018). The M4 Competition: Results, findings, conclusion and way forward. *International Journal of Forecasting 34*(4), 802–808.

Mazumdar, S. and S. Hoa (1995). Application of a Taguchi method for process enhancement of an online consolidation technique. *Composites 26*, 669–673.

Morris, R. and E. Watson (1998). A comparison of the techniques used to evaluate the measurement process. *Quality Engineering 11*, 213–219.

Mortimore, P., P. Sammons, L. Stoll, D. Lewis, and R. Ecob (1988). *School Matters*. Wells: Open Books.

Partridge, L. and M. Farquhar (1981). Sexual activity and the lifespan of male fruitflies. *Nature 294*, 580–581.

Raftery, A. E., D. Madigan, and J. A. Hoeting (1997). Bayesian model averaging for linear regression models. *Journal of the American Statistical Association 92*(437), 179–191.

Raghunathan, T. (2015). *Missing Data Analysis in Practice*. Boca Raton, FL: Chapman & Hall.

Rodriguez, N., S. Ryan, H. V. Kemp, and D. Foy (1997). Post-traumatic stress disorder in adult female survivors of childhood sexual abuse: A comparison study. *Journal of Consulting and Clinical Pyschology 65*, 53–59.

Rousseeuw, P. and A. Leroy (1987). *Robust Regression and Outlier Detection*. New York: Wiley.

Rousseeuw, P. J. and K. V. Driessen (1999). A fast algorithm for the minimum covariance determinant estimator. *Technometrics 41*(3), 212–223.

Schafer, J. (1997). *Analysis of Incomplete Multivariate Data*. London: Chapman & Hall.

Scheffé, H. (1959). *The Analysis of Variance*. New York: Wiley.

Sen, A. and M. Srivastava (1990). *Regression Analysis: Theory, Methods and Applications*. New York: Springer-Verlag.

Simonoff, J. (1996). *Smoothing Methods in Statistics*. New York: Springer–Verlag.

Steel, R. G. and J. Torrie (1980). *Principles and Procedures of Statistics, a Biometrical Approach*. (2nd ed.). New York: McGraw-Hill.

Steyerberg, E. W. (2009). *Clinical prediction models*, Volume 381. New York: Springer.

Stigler, S. (1986). *The History of Statistics*. Cambridge, MA: Belknap Press.

Stolarski, R., A. Krueger, M. Schoeberl, R. McPeters, P. Newman, and J. Alpert (1986). *Nimbus 7* satellite measurements of the springtime Antarctic ozone decrease. *Nature 322*, 808–811.

Thodberg, H. H. (1993). Ace of Bayes: Application of neural networks with pruning. Technical Report 1132E, Maglegaardvej 2, DK-4000 Roskilde, Denmark.

Tibshirani, R. (1996). Regression shrinkage and selection via the lasso. *Journal of the Royal Statistical Society, Series B 58*, 267–288.

Tukey, J. (1977). *Exploratory Data Analysis*. Reading, MA: Addison-Wesley.

Weisberg, S. (1985). *Applied Linear Regression* (2nd ed.). New York: Wiley.

Index

added variable plot, 93
additive model, 150
AIC, *see* Akaike information criterion
Akaike information criterion, 160
analysis of covariance, 7, 221
analysis of variance, 7, 241
ANCOVA, *see* analysis of covariance
ANOVA, *see* analysis of variance
autoregression, 56

B-splines, 148
back-transformation, 136
backward elimination, 156
balance, 64
balanced incomplete block design, 282
bias, 102
bias-variance tradeoff, 160
BIC, 161
blocking, 273
BLUE, 24
Bonferroni correction, 89, 233, 249
bootstrap, 48
Box–Cox method, 136
boxplot, 243
breakdown point, 128, 131
broken stick regression, 140

Cauchy distribution, 81
causal inference, 63
censoring, 211
centering, 107
central composite design, 32
central limit theorem, 44, 81, 97, 203
Choleski decomposition, 115
coefficient of determination, 26
collinearity, 108–112
complete case analysis, 213
completely randomized design, 273

condition number, 109
confidence intervals, 46–47, 53
confounding, 66, 68
contrast matrix, 235
controlling, 70
Cook statistics, 91, 205
correlated errors, 83–84
counterfactual, 63
crossvalidation, 183

datasets
    abrasion, 279
    air, 56
    cars, 103
    chmiss, 212
    chredlin, 199, 213
    coagulation, 242
    composite, 253
    corrosion, 123
    eco, 198
    ethanol, 142
    fat, 54
    fpe, 118
    fruitfly, 230
    galapagos, 20, 40, 61, 80, 137
    gala, 126
    globwarm, 83, 116
    jsp, 247
    manilius, 7
    meatspec, 179
    newhamp, 66
    oatvar, 274
    odor, 32
    pima, 3
    pvc, 257
    rabbit, 283
    savings, 77, 106, 137

295

seatpos, 110
sexab, 221
speedo, 268
star, 90, 128
state, 157
whiteside, 228
degrees of freedom, 18
dependent, 6
designed experiment, 64–65, 102
diagnostics, 75–99
dummy variable, 222
Durbin–Watson test, 84

ecological correlation, 197
effects, 241
eigenvalues, 109
Einstein, 1
endogenous, 6
equal variance, 75
errors, 15
errors in variables, 101
estimable function, 24
exogenous, 6
explanatory variable, 6
extrapolation, 55

F-test, 38
factor, 221
factorial, 266
factorial design, 253
false discovery rate, 249
familywise error rate, 249
feature selection, 155
features, 155
finite population, 46
fitted values, 18
fixed effects, 241
forward selection, 157
fractional factorial, 267, 279
future observation, 53

Gauss–Markov theorem, 24–25
generalized least squares, 84, 115
goodness of fit, 82, 121

half-normal plot, 85, 92

hat matrix, 18, 75, 85
heteroscedasticity, 76
hierarchical models, 156
hierarchy principle, 146
histogram, 81
history, 7–11
hockey-stick function, 141
Huber estimate, 125
hypothesis testing, 38

identifiability, 28–31
imputation, 215
incomplete blocks, 273
independent variable, 6
indicator variable, see dummy variable
influence, 91–93
initial data analysis, 2–6, 199
input, 6
interaction, 156, 225
intercept, 15–17

jackknife residuals, 88
jittering, 245

knotpoint, 141

lack of fit test, 121
lasso, 190
Latin square, 267, 278
least absolute deviation, 125
least squares, 18
level, 241
Levene's test, 245
leverage, 85
likelihood ratio test, 38, 137
linear dependence, 29
linear model, 15
logistic regression, 7
logit, 216
lognormal distribution, 81
lowess, 121

M-estimates, 125
Mahalanobis distance, 176
Mahalanobis distance, 85
matching, 67

matrix notation, 16
measurement error, 101
missing not at random, 212
missing at random, 212
missing completely at random, 211
missing values, 4, 211
model selection, 155
multicollinearity, *see* collinearity
multiple imputation, 217
multiple regression, 6
multiplicative errors, 135

natural experiment, 72
nominal factor, 256
nonconstant variance, 76
nonlinear models, 15
nonrandom sample, 45
normal equation, 18
normality, 80

observational data, 65–67
Occam's Razor, 155
offset, 43
ordered factor, 256
ordinal factor, 256
orthogonal projection, 17
orthogonality, 31, 173
outcome, 6
outlier, 81, 87–91
output, 6
overfitting, 124

p-value, 39
pairwise comparisons, 246
parameters, 15
partial correlation, 107
partial least squares, 184
partial regression, 93
partial residual plot, 94
Pearson residual, 86
penalized regression, 187
permutation test, 44–46
piecewise linear regression, 140
PLS, *see* partial least squares
Poisson distribution, 80
polynomial regression, 142

predicted values, 18
prediction, 53
predictor, 6
principal components, 173
projection, 18
pure error, 122

Q–Q plot, 80
QR decomposition, 22
quadratic model, 121
quantile regression, 128

random effects, 241, 261
randomization, 67, 273
randomized block design, 274
rank, 24, 39
reference level, 224
regression, 6
regression splines, 147
regression to the mean, 9
regularization, 187
relative efficiency, 278, 281
repeated measures, 122
replication, 122, 253
representative sample, 46
resampling, 48
residual sum of squares, 18
residuals, 17, 18, 88
response, 6
response surface, 146
ridge regression, 112, 187
robust regression, 81, 125–131
rounding error, 29
R^2, 26

sample of convenience, 46
sampling, 45
saturated model, 29
scale change, 105
segmented regression, 142
serial correlation, 132
Shapiro–Wilk test, 82
shrinkage, 182
SIMEX, 103
simple regression, 6
simulation, 48

singular, 28, 108
smoothing spline, 150
splines, 147
standardized residuals, 86
stepwise regression, 157
studentized residuals, 88
subspace, 42
sum coding, 236
swarmplot, 243

t-statistic, 41
t-test, 41, 222
testing sample, 179
Theil-Sen estimator, 129
time series, 56
training sample, 179
transformation, 79, 135–152, 203
treatment coding, 235
Tukey HSD, 247, 261
Tukey's non-additivity test, 255, 277

uniform distribution, 81

variable selection, 155–169
variance stabilization, 79
variance inflation factor, 109

weighted least squares, 117, 126
weights, 117, 227

Printed in the United States
by Baker & Taylor Publisher Services